Internet of Things in Automotive Industries and Road Safety

Electronic Circuits, Program Coding and Cloud Servers

RIVER PUBLISHERS SERIES IN TRANSPORT TECHNOLOGY

Indexing: All books published in this series are submitted to the Web of Science Book Citation Index (BkCI), to CrossRef and to Google Scholar.

The "River Publishers Series in Transport Technology" is a series of comprehensive academic and professional books which focus on theory and applications in the various disciplines within Transport Technology, namely Automotive and Aerospace. The series will serve as a multi-disciplinary resource linking Transport Technology with society. The book series fulfils the rapidly growing worldwide interest in these areas.

Books published in the series include research monographs, edited volumes, handbooks and textbooks. The books provide professionals, researchers, educators, and advanced students in the field with an invaluable insight into the latest research and developments.

Topics covered in the series include, but are by no means restricted to the following:

- Automotive
- Aerodynamics
- Aerospace Engineering
- Aeronautics
- Multifunctional Materials
- Structural Mechanics

For a list of other books in this series, visit www.riverpublishers.com

Internet of Things in Automotive Industries and Road Safety

Electronic Circuits, Program Coding and Cloud Servers

Rajesh Singh

Lovely Professional University
India

Anita Gehlot

Lovely Professional University
India

Raghuveer Chimata

Argonne National Laboratory
USA

Bhupendra Singh

Schematics Microelectronics
Dehradun, India

P. S. Ranjit

Aditya Engineering College
Andhra Pradesh, India

Published, sold and distributed by:
River Publishers
Alsbjergvej 10
9260 Gistrup
Denmark

River Publishers
Lange Geer 44
2611 PW Delft
The Netherlands

Tel.: +4536995319[7]
www.riverpublishers.com

ISBN: 978-87-70220-10-1 (Hardback)
978-87-70220-09-5 (Ebook)

Contents

Preface

The aim of writing this book is to provide a platform to the readers, where they can access the applications of 'Internet of Things' in the field of automotive. The book provides the basic knowledge of the modules with their interfacing along with the programming

The objective of this book is to discuss various applications in automotive industries where 'Internet of things' can play important role. Few examples for rapid prototyping are included, to make the readers understand about the concept of IoT.

This book comprises of total ten chapters for designing different independent prototypes for the automotive applications. It would be beneficial for the people who want to get started with hardware based project prototypes.

This book is entirely based on the practical experience of the authors while undergoing projects with the students and industries. We acknowledge the support from Nutty Engineer.com, to use its products to demonstrate and explain the working of the systems. We would like to thank the publisher for encouraging our idea about this book and the support to manage the project efficiently. Although the circuits and programs mentioned in the text are tested on real hardware but in case of any mistake we extend our sincere apologies. Any suggestions to improve the contents of the book are always welcome and will be appreciated and acknowledged.

ACKNOWLEDGEMENTS

We acknowledge the support from Nutty Engineer to use its products to demonstrate and explain the working of the systems. We would like to thank RIVER PUBLISHER for encouraging our idea about this book and the support to manage the project efficiently.

We are grateful to the honorable Chancellor (Lovely Professional University) Ashok Mittal, Mrs. Rashmi Mittal (Pro Chancellor, LPU), Dr. Ramesh Kanwar (Vice Chancellor, LPU), Dr. Loviraj Gupta (Executive Dean, LPU) for their support. We are also thankful to the chancellor (UPES) Dr. S. J. Chopra, Dr. Dependra Jha (Vice Chancellor, UPES), Dr. Kamal Bansal (Dean, SoE,

UPES), Dr. Piyush Kuchhal (Associate Dean, UPES) and Dr. Suresh Kumar (Director, UPES) for their support and constant encouragement. In addition we are thankful to our family, friends, relatives, colleagues and students for their moral support and blessings.

Dr. Rajesh Singh
Dr. Anita Gehlot
Dr. Raghuveer Chimata
Mr. Bhupendra Singh
Dr. P. S. Ranjit

List of Figures

List of Tables

List of Abbreviations

API	Application program interface
APP	Application
ERP	Enterprise source planning
GPRS	General Packet Radio Service
GPS	Global Positioning System
HTTP	Hyper Text Transfer Protocol
IDE	Integrated development environment
IIoT	Industrial Internet of Things
IoT	Internet of Things
IPv6	Internet Protocol version 6
LCD	Liquid crystal display
LED	Light Emitting Diode
MISO	Master In Slave Out
MOSI	Master Out Slave In
OSI	Open Systems Interconnection
PC	Personal Computer
REST	Representational state transfer
RF	Radio Frequency
RFID	Radio frequency identification
TCP	Transmission Control Protocol
UART	Universal synchronous and asynchronous receiver-transmitter
V2V	Vehicle-to-vehicle
WSN	Wireless sensor networks

1

Introduction

1.1 Introduction to IoT

As per Cambridge English Dictionary, the meaning of "Internet" is the large systems of connected computers around the world that allows people to share information and communicate with each other and Things means "used to refer in an approximate ways to objects."

Internet of things (IoT) is the communication between abiotic without interference of biotic systems, i.e., any product/process/service can interact within them and with other products or processes without interference of human beings.

Few examples:

1. Operate AC/Geyser with a mobile.
2. Automatic gate opening of a house.
3. Authorize a person to enter a home in the absence of the owner.
4. Health monitoring system.
5. An automatic interacting system with a driver to avoid the accidents.

The required data can be collected by the IoT-enabled devices from the existing wide variety of technologies and send the data as and when required to their identified devices. In present market, smart ACs and heaters act according to the requirement of the user, using a WiFi system. Even our cutlery will start informing us about the quantity of food intake as per calories' requirement; further, it may depend on BMI and the type of workout. The old mobile phones, TVs, and house dustbins are becoming smarter. As per one statistical analysis, IoT-enabled devices will reach upto 31 billion in numbers by 2020. IPv6 is version of the Internet protocol, which provides the identification for computers on network and routes the traffic over the Internet. IPv6 has 128-bit Hex numbers address.

IoT – Key Features

The basic architecture of IoT comprises sensors, actuators, and their enabling machine language. Artificial intelligence, its connectivity, and its active engagement can be used by small devices.

Artificial Intelligence: Artificial intelligence is a mathematically developed manmade machine intelligence, developed in order to perceive the natural environment to achieve the target. The IoT-enabled artificial intelligence has the smart algorithm to collect the data and self-communicating among connected devices through their networks. For example, in a smart bin system of a production line, if the material gets over, then data will be transferred to the enterprise resource planning (ERP) system, followed by order received by the supplier from the ERP and refills the intelligent smart bin.

Connectivity: Connectivity is a major issue in most of the places. Earlier the industries had connectivity. At present, the XBee, RFID, RX/TX 433MHz, and WiFi are the devices to provide network connectivity to realize the IoT applications.

Sensors: A device in the form of sensors/transducers is required to detect the physical parameters and communicate its data to the destination through the embedded system.

Active Engagement: The IoT being an active engagement of technologies makes a paradigm shift over today's passive engagement among service and product managements.

Small devices: The IoT is a small device, which enables and ensures more precision, scalability, and versatility.

IoT Merits

1. In the today scenario, the IoT becomes a part of personal as well as business life of an individual.
2. It improves the customer engagement with product service.
3. It is an optimized way to use the technology.
4. The data collection is easy.

IoT – Demerits

Though IoT addressed so many meritorious things, it also has some challenges as well:

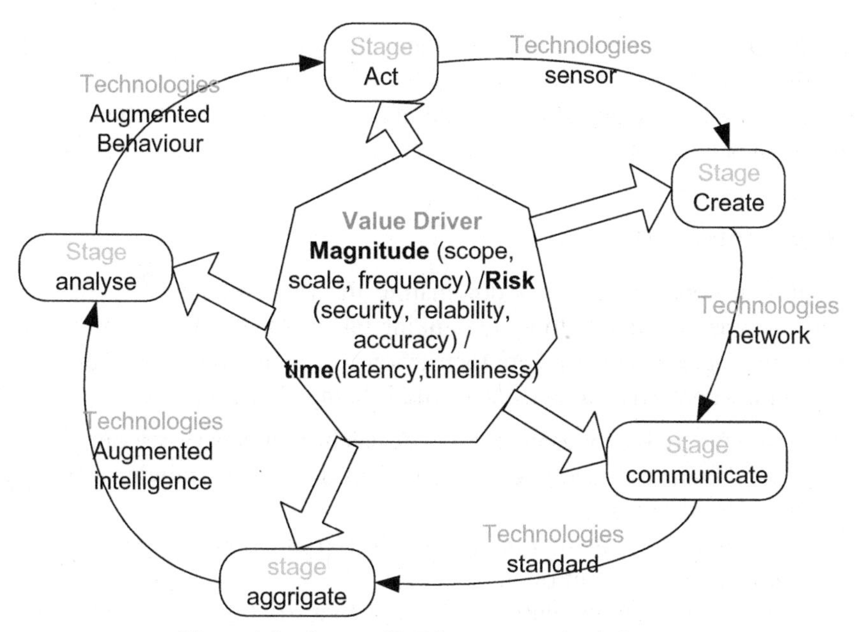

Figure 1.1 Stages of IoT in an automotive industry.

1. Security: In these days, everyone is listening to the word "Cyber security" because of each individual communicating with each other through virtual networks, which becomes an advantage for hackers.
2. Complexity: To make systems/processes more simpler and user friendly, the complexity of developing them always increases.
3. Compliance: Any service or technology in the real business needs to be comply with regulations.

The IoT is the integration of sensing, communication, and its analytic capabilities raised overall conventional technologies. The IoT area promises in helping automotive industries by directly managing their existing assets at different places, supply chains, and after sales service, dealers and customer relationships, which helps to understand and access the data/information as and when required. Figure 1.1 shows the stages of IoT in automotive industry.

1.2 Future and Market Potential of IoT

Society is becoming smart by using the IoT eco-system in day-to-day life activities, which leads to enrich the lives of human beings. Its application started from a smart dustbin to operate the parking garage door. In the near future, IoT can become multi-trillion dollars' industry. CISCO revealed that the use of IoT-enabled devices number may raise to 50 billion against 7.6 billion people [1].

1.3 Industry 4.0

The present conceptual age is dominating the right brained activities like creativity when compared to logic thinking taken place by the left side brain information age, as industrial era focused on change in every aspect of life to digital era, where every dimension of life is being changing.

1. First industrial revolution (1784) – A mechanical weaving loom.
2. Second industrial revolution (1923) – Introduction of a moving assembly line at Ford Motors.
3. Third industrial revolution (1969) – Introduction of the first programmable logic controller.
4. Fourth industrial revolution (2014) – Industry4.0 was introduced to intelligent machines, embedded cyber physical sensors, collaborative technologies, and network processes [2].

The term Industry 4.0 was used by the German Government, which means the use of IoT in the manufacturing industry. The term Industry 4.0 refers to the fourth industrial revolution. Sometimes it is also known as **Industrial** IoT **(IIoT)** [11].

Industry 4.0 is based on the technological concepts of cyber-physical systems, the IoT, and the Internet of Services. It facilitates the vision of the Smart Factory. IoT focuses on convenience for individual consumers, whereas Industrial 4.0 is strongly focused on improving the efficiency, safety, and productivity of operations with a focus on return on investment. Although Industry 4.0 is more particular to industry, yet the two terms refer to similar movements. Industry 4.0 represents a paradigm shift from "centralized" to "decentralized" smart manufacturing. The BMW adopted Industry 4.0 manufacturing strategy in order to obtain their greater efficiency and flexibility in their factories. Earlier, the automotive parts were assembled by robots and human beings (technicians). But most of the insignificant assembly tasks were performed by human hands, whereas with Industry 4.0, technicians/engineers are working with interactive robots. Figure 1.2 shows the four stages of industrialization and Figure 1.3 shows the Industry 4.0 with BMW.

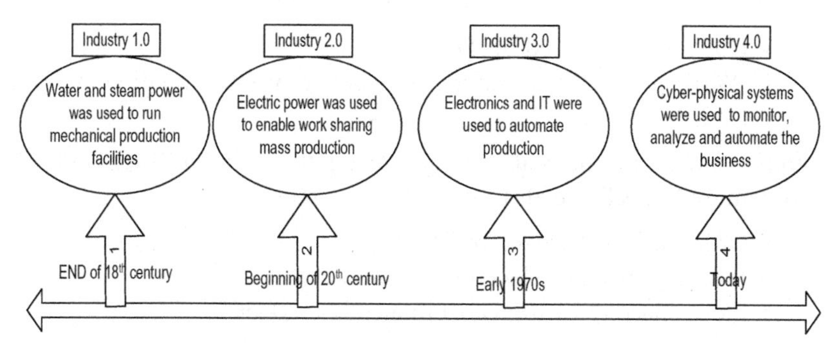

Figure 1.2 Four phases of industrialization.

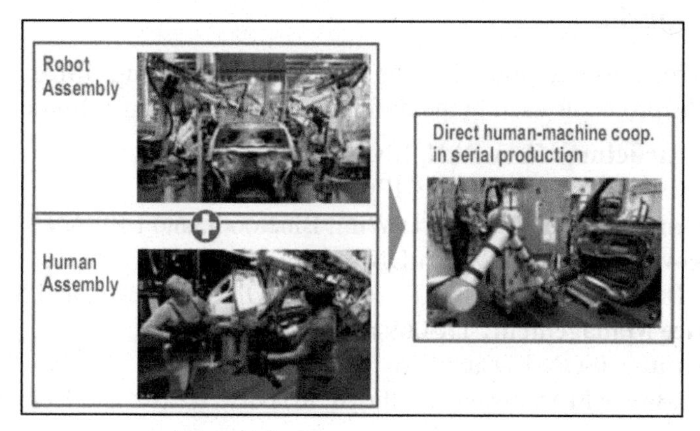

Figure 1.3 Industry 4.0 with BMW.

Advantages of Industry 4.0 application in Automotive Industry:

- Robots are having interactive nature required to work with humans, which makes them much safer and user friendly.
- These robots can eliminate hard physical labor being a part of human team, which leads to increase in production efficiency of the plant.
- Robot can use its energy and mechanical accuracy to support human work force in healthy conditions for long lasting.

Case studies:

1. Predictive maintenance executed by BOSCH, IBM, and US Department of Energy using connectivity and big data analytics leads to the following points:
    - Maintenance cost reduced by 30%
    - Unplanned breakdowns reduced by 75%

- Down time reduced by 45%
- Working condition, safety, and better product quality increased by 25%.

2. Improvement of internal logistics using the camera/RFID and ERP system by WURTH leads to the following points:
   - Ordering cost reduced by 80%
   - Working capital reduced by 20–40%.
   - Floor space utilization cost reduced by 20–30%.

1.4 IoT Model

The IoT protocols have the following layers like the existing OSI architecture model. The protocols used in the different categories are as follows:

1. **Infrastructure:** 6LowPAN, IPv4/IPv6, and RPL
2. **Identification:** EPC, uCode, IPv6, and URIs
3. **Communication/Transport:** WiFi, Bluetooth, and LPWAN
4. **Discovery:** Physical Web, mDNS, and DNS-SD
5. **The Data Protocols:** MQTT, CoAP, AMQP, Websocket, and Node
6. **Device Management:** TR-069 and OMA-DM
7. **Semantic:** JSON-LD and Web Thing Model
8. **Multi-layer Frameworks:** Alljoyn, IoTivity, Weave, and Homekit [2].

Figure 1.4 shows the IoT protocol architecture model for various applications.

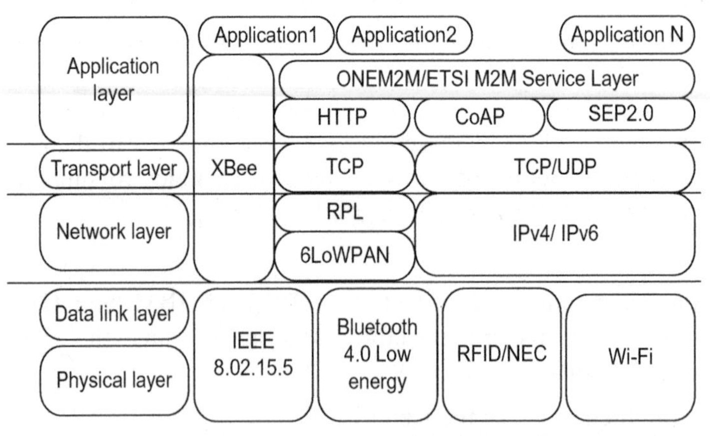

Figure 1.4 IoT protocol architecture model.

1.5 IoT Protocol Architecture

Figure 1.5 shows the architecture of IoT. It contains the perception layer, network layer, and application layer.

Perception Layer – In this layer, the collection of the data takes place from the environment like air pressure, altitude, temperature, and humidity from various homogeneous or heterogeneous devices. Intelligent sensors in wireless sensor networks (WSNs) collect and process the data for different applications. Devices like actuators linear/rotational, cameras, intelligent sensors, and GPS are communicating with each other through different protocols like XBee, 2.4GHz RF modem, 433MHz RF modem, Bluetooth, nRF modem, and WiFi for different ranges of communications.

Network Layer – The network layer establishes a link between the perception layer and the application layer. The network layer collects the information from the perception layer and sends it to the application layer for further processing.

Application Layer – The application layer bridges the gap between the application and users.

Figure 1.6 shows other ways of representing the architecture/communication protocol stack of IoT. The communication protocol stack of IoT has the

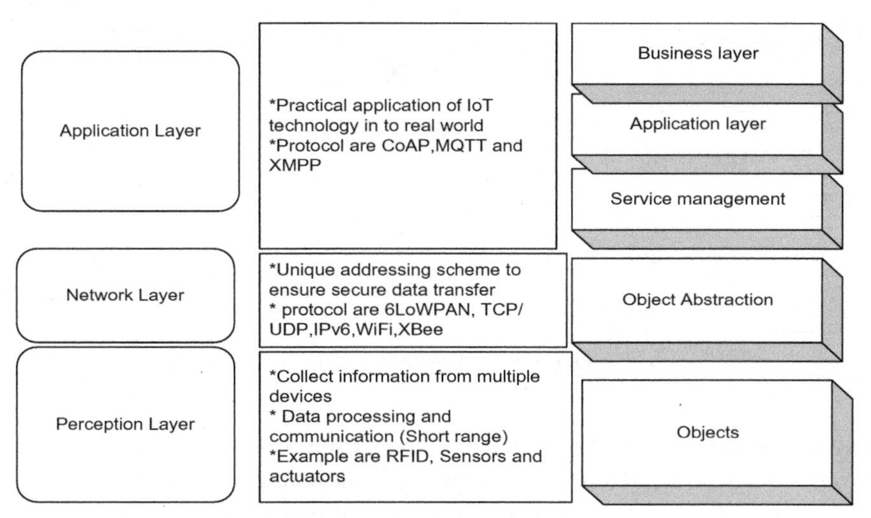

Figure 1.5 IoT architecture.

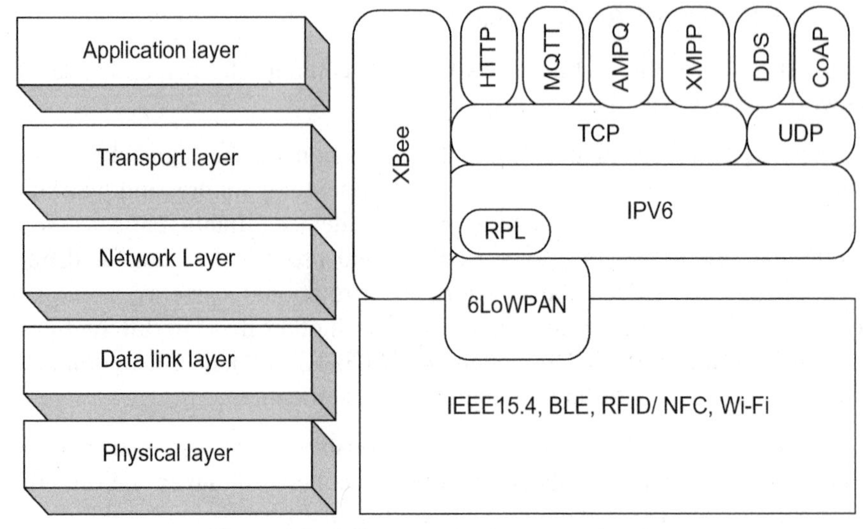

Figure 1.6 IoT communication protocol stack.

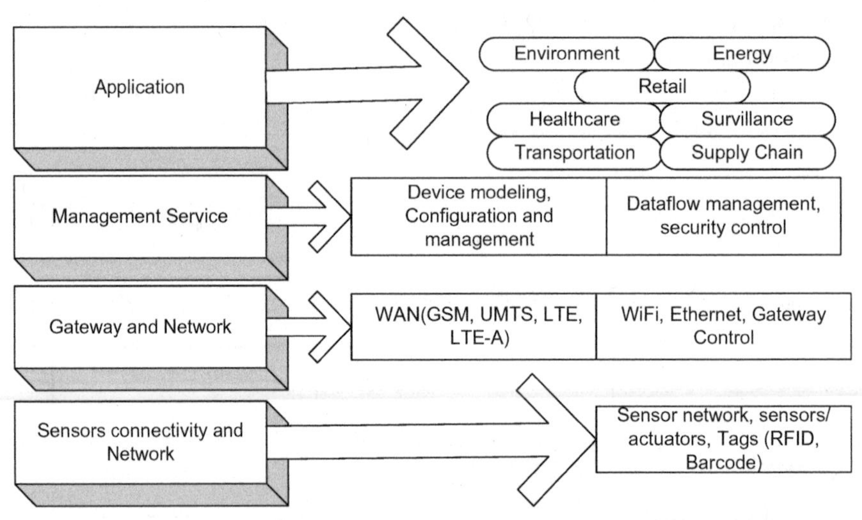

Figure 1.7 Architecture of IoT.

physical layer, data link layer, network layer, transport layer, and application layer with its associated protocol. Figure 1.7 shows the IoT architecture.

Figure 1.8 shows other ways of representing the architecture of IoT. The radio frequency identification device (RFID) tags and sensors are the important part of an IoT system and these are responsible for collecting the

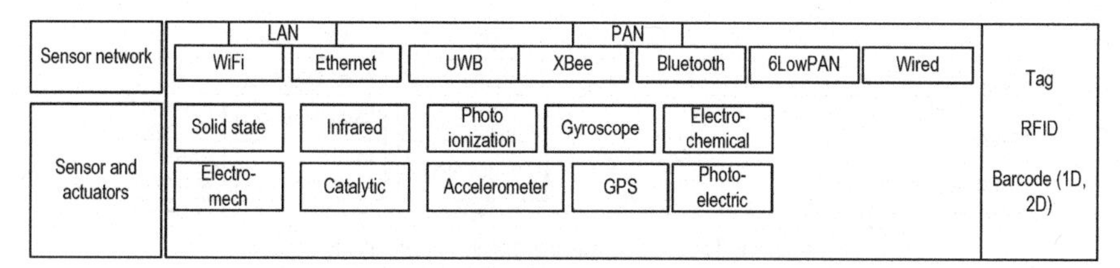

Figure 1.8 Sensor, connectivity, and network layer.

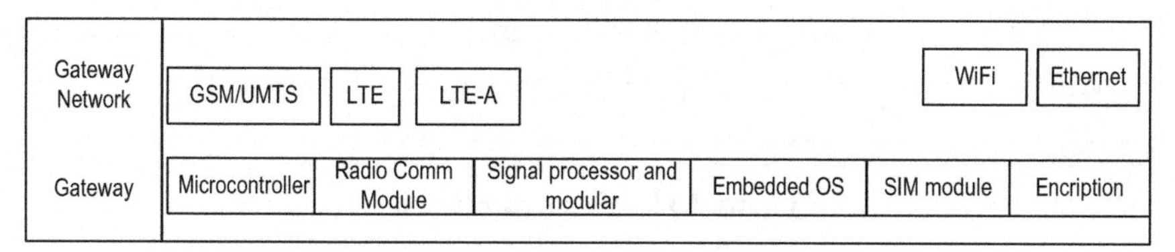

Figure 1.9 Gateway and network layer.

raw data. It has the sensor, connectivity, and the network layer at the bottom of this layer and has RFID tags or barcode reader, sensors/actuators, and then the communication networks.

Figure 1.9 shows the gateway and network layer. This layer is responsible for providing the route to the data, coming from the sensor, connectivity, and network layer and passes it to the next layer, which is the management service layer. This layer has large storage capacity to store the data of sensors, RFID tags, etc. This layer also integrates various network protocols as different IoT devices work on different kinds of network protocols. The gateway at the bottom of the Figure 1.9 contains embedded OS, signal processors, microcontrollers, etc. The gateway networks contain WiFi, Ethernet, local area network, wide area network (WAN), etc.

Figure 1.10 shows the management service layer used for managing the IoT services. The management service layer is responsible for securing analysis of IoT devices, analysis of information (stream analytics and data analytics), and device management. Data management is required to extract the necessary information from the enormous amount of raw data collected by the sensor devices to yield a valuable result of the collected data. The management service layer has operational support service, which includes device modeling, device configuration and management, and many more. A billing support system supports billing and reporting. IoT/M2M application service includes analytics platform and security which includes access

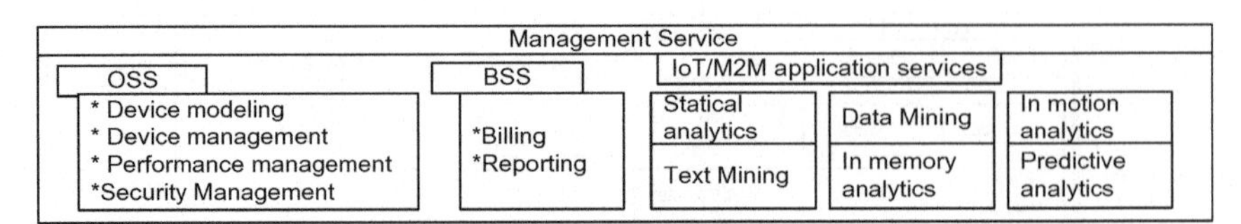

Figure 1.10 Management service layer.

Application management
Environment
Energy
Transporation
Healthcare
Retail

Fleet tracking
Asset management
Supply chain
People tracking
Survillance

Figure 1.11 Application layer.

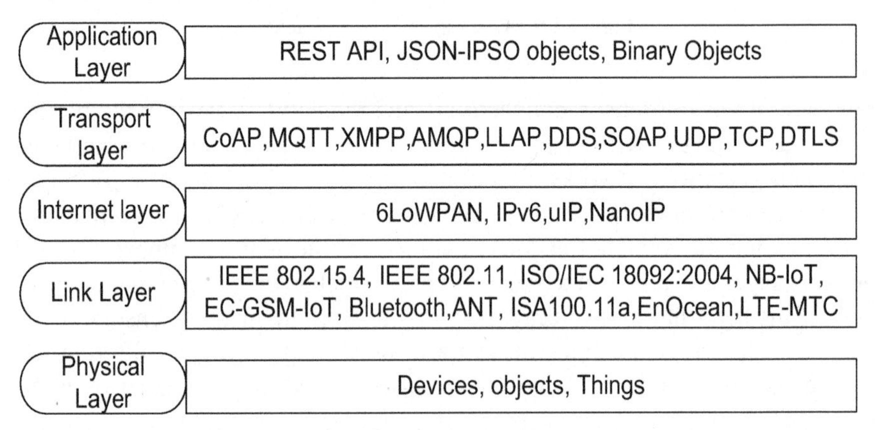

Figure 1.12 Protocol architecture.

controls, encryption, identity access management, business rule management, and business process management.

Figure 1.11 shows that the application layer forms the topmost layer of the IoT architecture, which is responsible for effective utilization of the data collected. The IoT applications include home automation, industrial automation, healthcare, transportation, surveillance, retail, and tracking.

Figure 1.12 shows the other way to represent the IoT protocol architecture [5].

Figure 1.13 shows the taxonomy of research in IoT technologies, where perception, preprocessing, communication, middleware, and application are the sequence of execution of the primary data collected from the field.

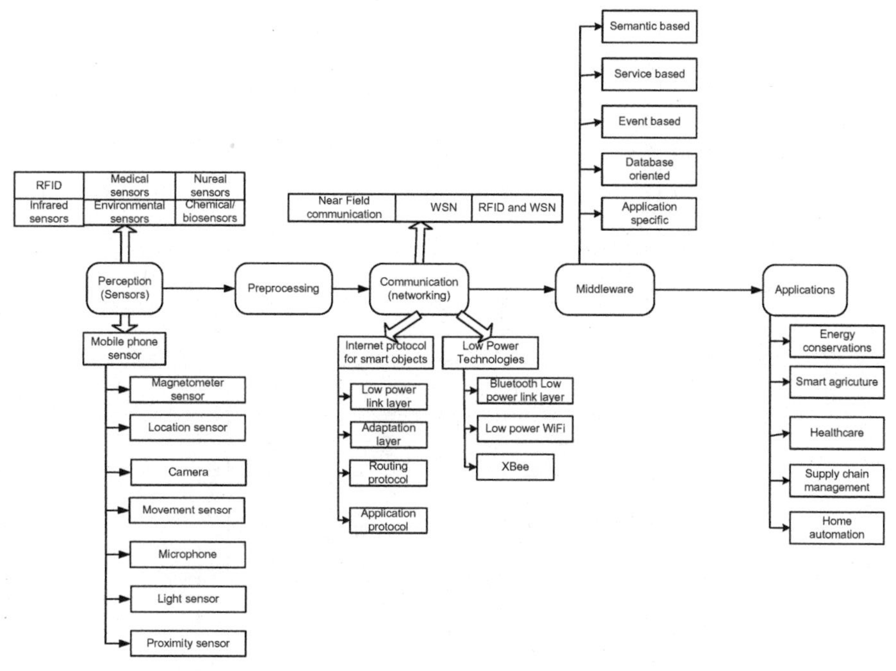

Figure 1.13 Taxonomy of research in IoT technologies [4].

1.6 IoT Technology

The IoT architecture consists of a collection of active devices/things, digital and analog sensors, linear and rotational actuators, communication protocols, and developers. Table 1.1 shows the enabling technologies for IoT. The enabling technologies like WiFi, WiMAX, LRWPAN, Bluetooth, and LoRA are differentiated with respect to its standard, frequency, data rate, transmission range, energy consumption, and cost.

Table 1.2 shows the IoT supported platform with its parameters. It shows IoT-enabled devices to control through Electric Imp 003 Raspberry Pi BC, Intel Galileo Gen 2, Intel Edison, Beagle Bone Black, Arduino Uno, Arduino Yun, ARM embed NXP LPC1768, and TelosB. These platforms have been identified by using parameters like general-purpose unit, clock, voltage requirement, Flash memory, system memory, integrated development environments (IDEs), programming languages, input and output connectivity, and type of processor.

Table 1.1 IoT technologies

Parameters	WiFi	WiMAX	LRWPAN	Mobile	Bluetooth	LoRA
Standard	IEEE802.11a/c/b/d/g/n	IEEE802.16	IEEE802.15.4 XBee	2G GSM, CDMA 3G-UMTS, CDMA2000, $G LTE	IEEE802.15.1	LoRA WAN R1.0
Frequency band	5–60 GHz	2–66 GHz	868/915 MHz, 2.4 GHz	865 MHz, 2.4 GHz	2.4 GHz	868/900 MHz
Data rate	1–6.75 Mb/s	1 Mb/s–1 Gb/s (fixed) 50Mb/s–100Mb/s (mobile)	40–250 kb/s	2G: 50–100 kb/s 3G: 200 kb/s 4G: 0.1–1 Gb/s	1–24 Mb/s	0.3–50 kb/s
Transmission range	20–100 m	<50 km	10–20 m	Entire cellular area	8–10 m	<30 km
Energy consumption	High	Medium	Low	Medium	Low	Very low
Cost	High	High	Low	Medium	Low	High

Table 1.2 IoT supported platform

Parameters	Arduino Uno	Arduino Yun	IntelGalileo gen2	Intel Edison	Beagle Bone Black	Electric IMP003	Raspberry Pi	ARM mbed NXP LPC1768	Telos B
Processor	Atmega 328p	Atmega32U4 and Atheros AR9331	Intel Quark SOC X1000	Intel Quark SOC X1000	Sitara AM3358BZC Z100	ARM cortex M4F	Broadcom BCM2835 Soc-based ARM11	ARM cortex M3	MSP430
Operating voltage	5 V	5, 3 V	5 V	3.3 V	3.3 V	3.3 V	5 V	5 V	3–3.6 V
Clock speed	16 MHz	16, 400 MHz	400 MHz	100 MHz	1 GHz	320 MHz	700 MHz	96 MHz	8 MHz
Bus width	8	32	32	32	32	32	32	32	16
System memory	2 kB	2, 5, 64 kb	256 MB	1 GB	512 MB	120 kb	512 MB	32 kb	10 kb
Flash memory	32 kb	32 kb, 16 Mb	8 Mb	4 GB	4 GB	4 Mb	–	512 kb	48 kB
EEPROM	1 kb	1 kb	8 kb	–	–	–	–	–	–
Development environment	Arduino IDE	Arduino IDE	Arduino IDE	Arduino IDE, Eclipse, Intel XDK	Debian, android, Ubantu, Cloud9 IDE	Electric ImpIDE	NooBS	C/C++ SDK, Online compiler	Eclipse IDE
Programming language	Wiring	Wiring	Wiring	Wiring, C, C++, Node JS, HTML5	C,C++, Python, Perl, Ruby, JAVA, Node8	Squirrel	C,C++, Python, Ruby, JAVA	C,C++	C, NesC
I/O connectivity	SPI, I2C, UART, GPIO	SPI, I2C, UART, GPIO	SPI, I2C, UART, GPIO	SPI, I2C, UART, GPIO, I2S	SPI, I2C, UART, GPIO, McASP	SPI, I2C, UART, GPIO	SPI, DSI, UART, SDIO, CSI, GPIO	SPI, I2C, GPIO, CAN	USB serial, GPIO
Communication standard	IEEE 802.11 b/g/n, IEEE 802.15.4, 433RF, BLE4.0, Ethernet, Serial	IEEE 802.11 b/g/n, IEEE 802.15.4, 433RF, BLE4.0, Ethernet, Serial	IEEE 802.11 b/g/n, IEEE 802.15.4, 433RF, BLE4.0, Ethernet, Serial	IEEE 802.11 b/g/n, IEEE 802.15.4, 433RF, BLE4.0, Ethernet, Serial	IEEE 802.11 b/g/n, IEEE 802.15.4, 433RF, BLE4.0, Ethernet Serial	IEEE 802.11 b/g/n, IEEE 802.15.4, 433RF, BLE4.0 Ethernet, Serial	IEEE 802.11 b/g/n, IEEE 802.15.4, 433RF BLE4.0, Ethernet, Serial	IEEE 802.11 b/g/n, IEEE 802.15.4, 433RF BLE4.0, Ethernet, Serial	CC2420

1.7 Functional Block of IoT

An IoT system comprises many functional blocks to facilitate various utilities to the system such as sensing, identification, actuation, communication, and management.

Figure 1.14 shows the various components of IoT devices like connectivity, processors, audio/video interfaces, input–output interfaces, storage interface, memory interface, and graphics. Figure 1.15 shows the functional blocks of IoT, which consists of physical devices, communication modem, service, security, management, and application.

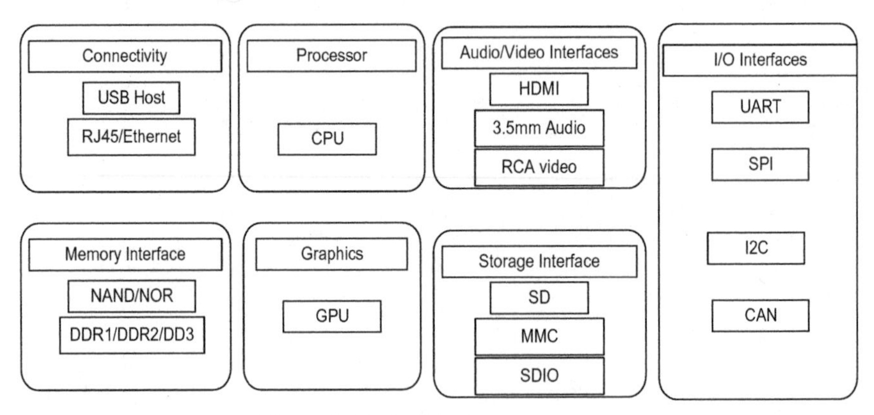

Figure 1.14 IoT device components.

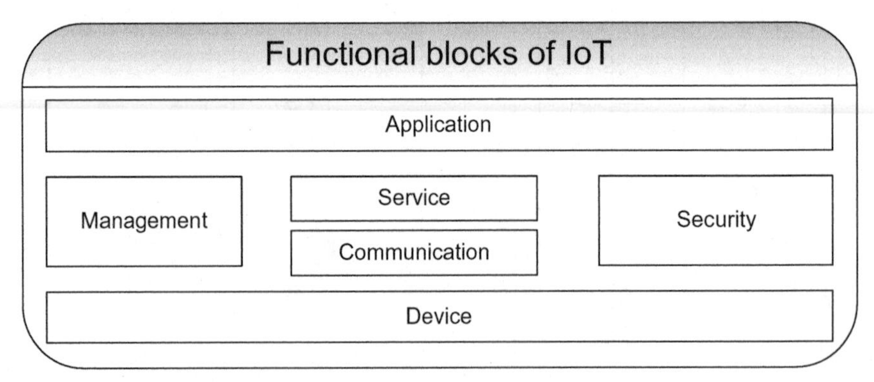

Figure 1.15 Functional blocks of IoT.

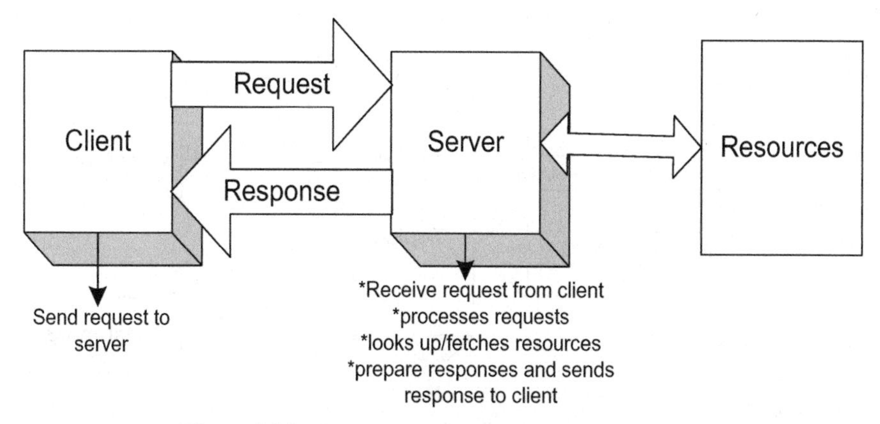

Figure 1.16 Request–response communication model.

1.8 IoT Communication Models

The communication model of IoT includes request–response communication model, publish subscribe communication model, push–pull communication model, and exclusive pair communication model.

1.8.1 Request–Response Communication Model

Figure 1.16 shows the request–response communication model. The request response is a communication model in which the client sends a request to the server and the server responds to the request. When the server receives a request, it decides how to respond, fetches the data, retrieves resource representations, prepares the response, and then sends the response to the client.

1.8.2 Publish Subscribe Communication Model

Figure 1.17 shows the block diagram of the publish subscribe communication model. It includes consumers, consultants, and publishers. Publishers are source of data availability; they send the same to the aspirants through consultants as and when requirement raises.

1.8.3 Push–Pull Communication Model

Figure 1.18 shows the push–pull communication model of IoT. Queue plays the important role in this push–pull communication model. Data is pushed

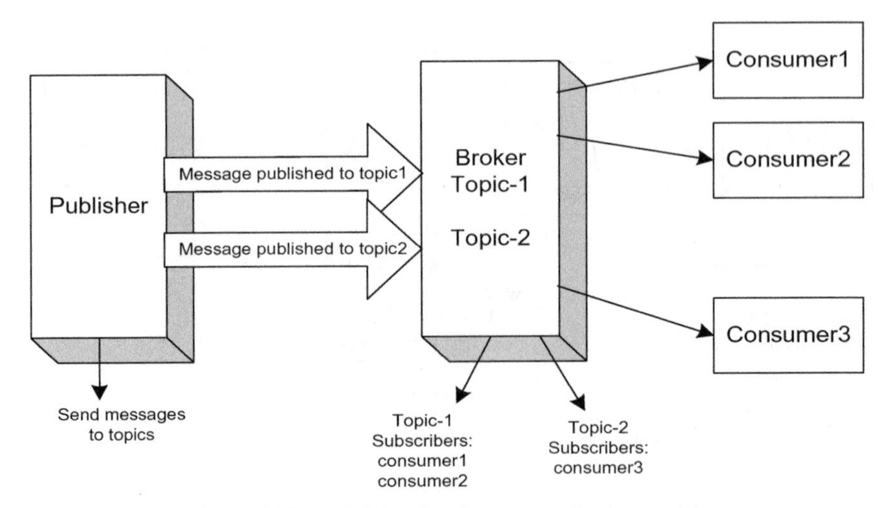

Figure 1.17 Publish subscribe communication model.

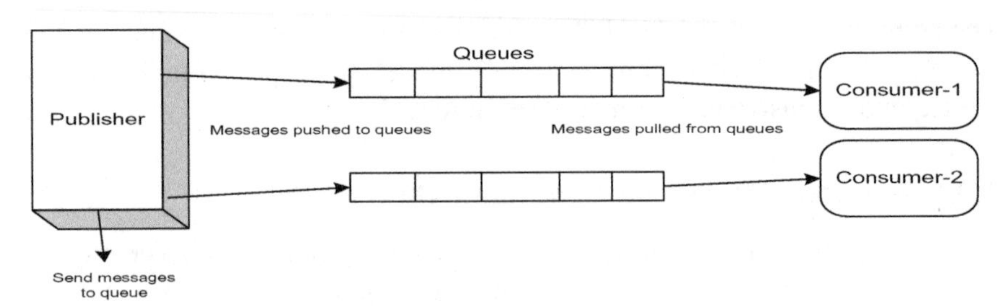

Figure 1.18 Push–pull communication model.

to the queue and pulls back when needs to consume. Queue is helping to communicate between the manufacturer and the consumer. If there is any mismatch in between, push and pull systems can be avoided and also they are acting as a buffer.

Figure 1.19 shows the exclusive pair model. It is a bi-directional, full-duplex communication model that uses a persistent connection between the client and the server. Once the connection is set up, it remains open until the client sends the request to close the connection. The client and the server can send the messages to each other after the connection setup.

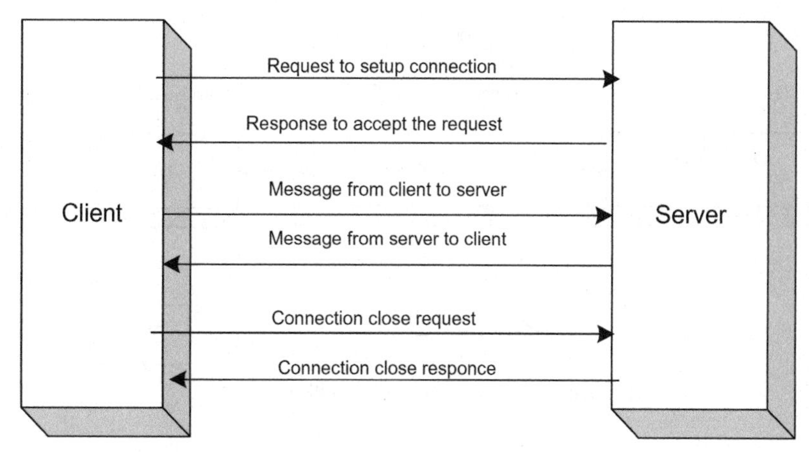

Figure 1.19 Exclusive pair model.

1.9 IoT Communication API

The most popular IoT communication application programmable interfaces (APIs) are representational state transfer (REST) and WebSocket.

1.9.1 REST-based Communication API

The REST or REST services are the way to provide interoperability between computer systems connected to the Internet. Such a type of web service provides a request to the system to access and manipulate the textual representations of web resources. The system uses the uniform and predefined set of operations, which expose the arbitrary sets of operations such as WSDL and SOAP. The web resources are available in World Wide Web (www) as documents or files identified by their URLs. Figure 1.20 shows the communication with REST API.

Figure 1.21 shows the request–response model of RSET. The request–response is a communication model in which the client sends a request to the server and the server responds to the request. When the server receives a request, it decides how to respond, fetches the data, retrieves resource representations, prepares the response, and then sends the response to the client.

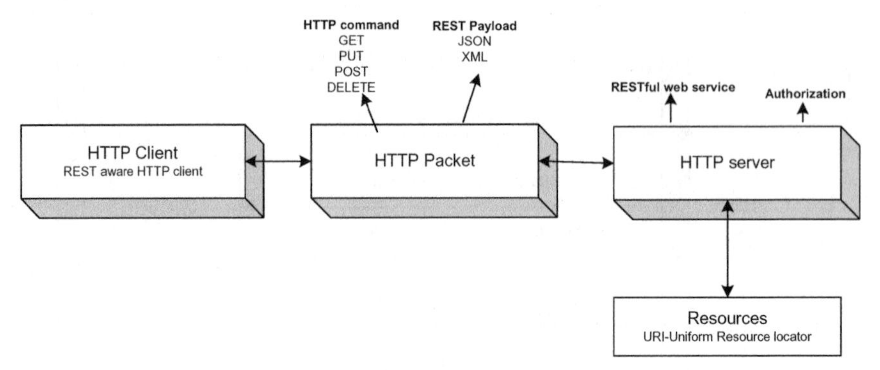

Figure 1.20 Communication with REST API.

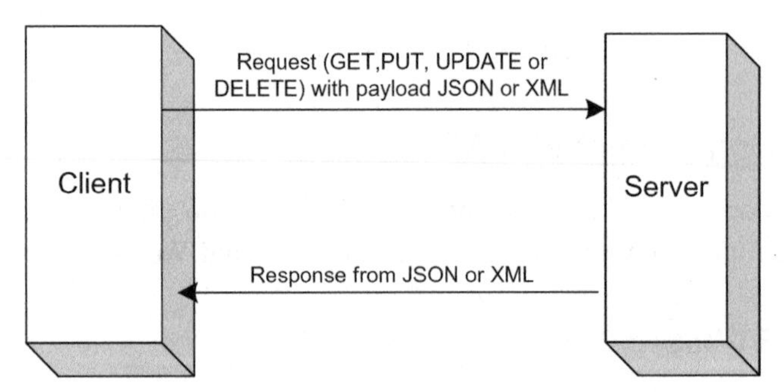

Figure 1.21 Request–response model of RSET.

1.9.2 WebSocket-based Communication API

WebSocket is a computer communication protocol, which provides full-duplex communication channels using a single TCP connection. The Web-Socket is application program interface (API) in Web IDL and standardized by the W3C. It is a two-way (bi-directional) conversation and able to establish the communication between the client and the server.

Figure 1.22 shows the exclusive pair model used by WebSocket APIs. The exclusive pair is a bidirectional, full-duplex communication model that uses a persistent connection between the client and the server. Once the connection is set up, it remains open until the client sends the request to close the connection. The client and the server can send the messages to each other after the connection setup. In the exclusive pair communication model, the server is aware of all the open connections.

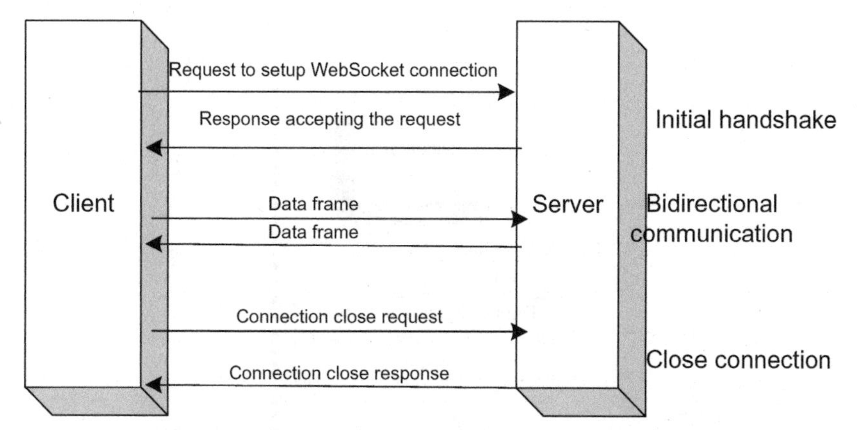

Figure 1.22 Exclusive pair model used by WebSocket APIs.

1.10 IoT Levels

The IoT levels can be classified into six categories. The details are as follows.

1.10.1 Level-1 IoT System

A level-1 IoT system has a single node/device that performs sensing operation and/or actuation, stores data, performs analysis, and hosts the application as shown in Figure 1.23. The system is suitable for modeling low-cost and low-complexity solutions where the data involved are not big and the analysis requirements are not computationally intensive. Home automation is the best example, where a single node can control the lights and appliances in a home remotely.

1.10.2 Level-2 IoT System

A level-2 IoT system has a single node/device that performs sensing and/or actuation and local analysis as shown in Figure 1.24. The data are stored in the cloud and the application is cloud based. The system is suitable where the data involved are big. However, the primary analysis requirement is not computationally intensive and can be done locally itself. Smart irrigation is an example of this level; in this system, a single node monitors the soil moisture level and controls the irrigation system.

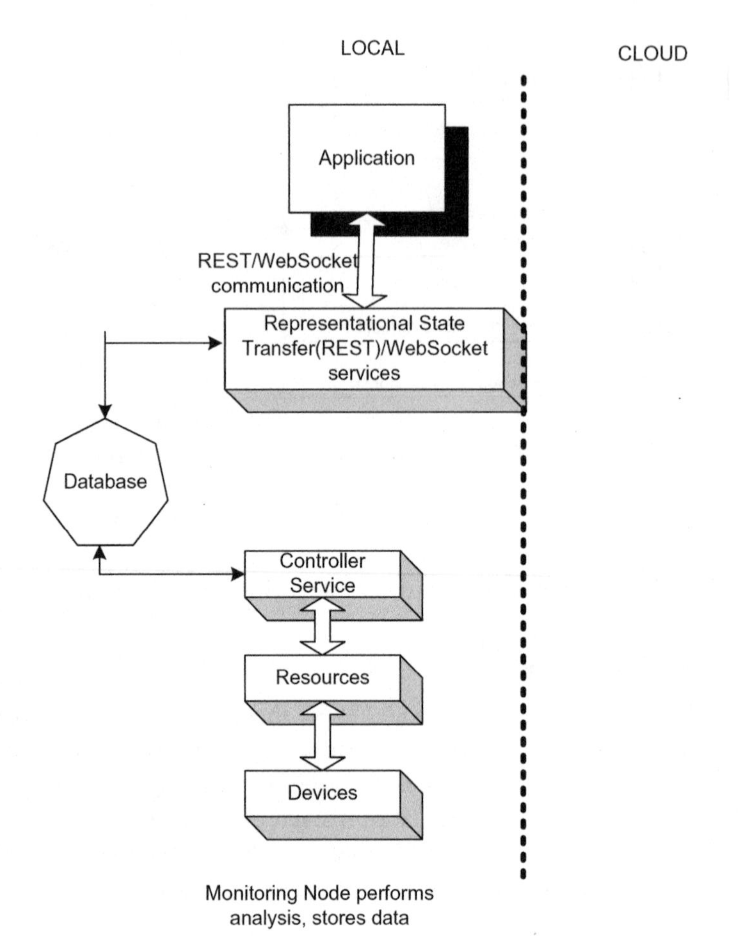

Figure 1.23 IoT level-1.

1.10.3 Level-3 IoT System

A level-3 IoT system has a single node, where data are stored and analyzed in the cloud and application is cloud based as shown in Figure 1.25. This type of system is suitable where data involved are big and the analysis requirements are computationally intensive. Tracking of package is an example for this, where a single node at one place always monitors the coordinates of the package being supplied.

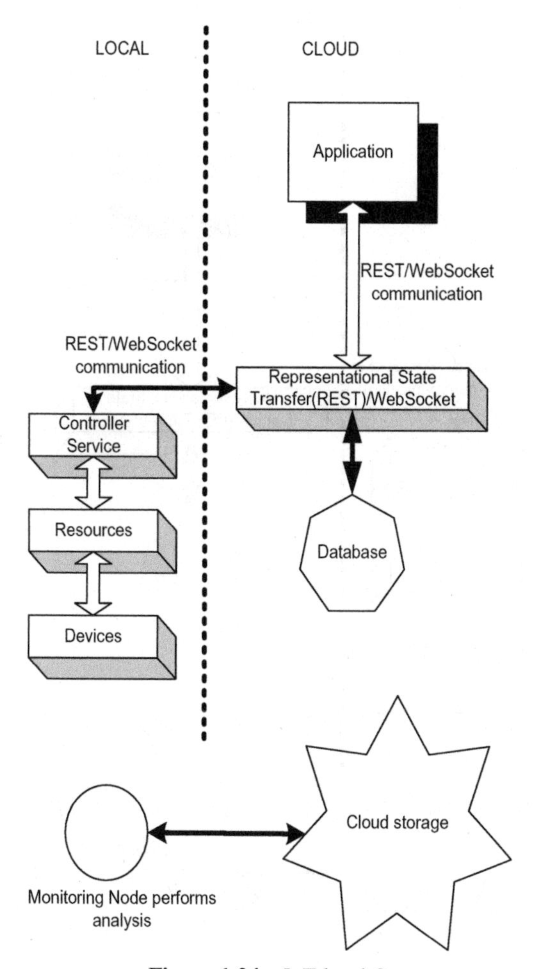

Figure 1.24 IoT level-2.

1.10.4 Level-4 IoT System

A level-4 IoT system has multiple nodes that perform local analysis (Figure 1.26). Data are stored in the cloud. The local and cloud-based observer nodes can subscribe to and receive information collected in the cloud from IoT devices. Observer nodes can process information and use it for various applications. However, observer nodes do not perform any control functions. These types of systems are suitable where a lot of nodes are involved in the process and the data are big. The analysis requirements are computationally intensive. The city noise monitoring is one best example in

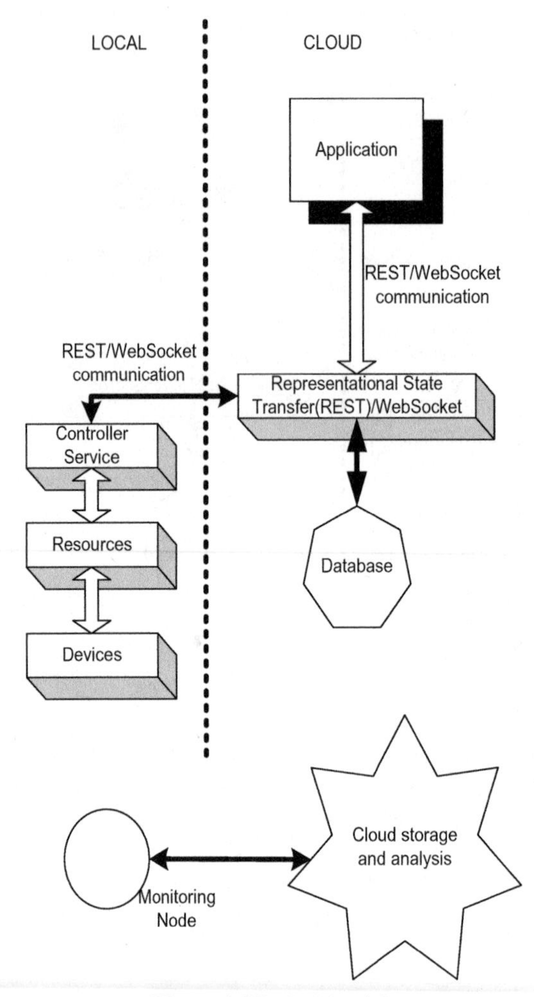

Figure 1.25 IoT level-3.

the level. The system consists of multiple nodes placed in different locations for monitoring noise levels in an area.

1.10.5 Level-5 IoT System

A level-5 IoT system has a large number of end nodes and one coordinator (Figure 1.27). The end node performs the sensing and the coordinator node collects data from the end nodes and sends it to the cloud. Data are stored and

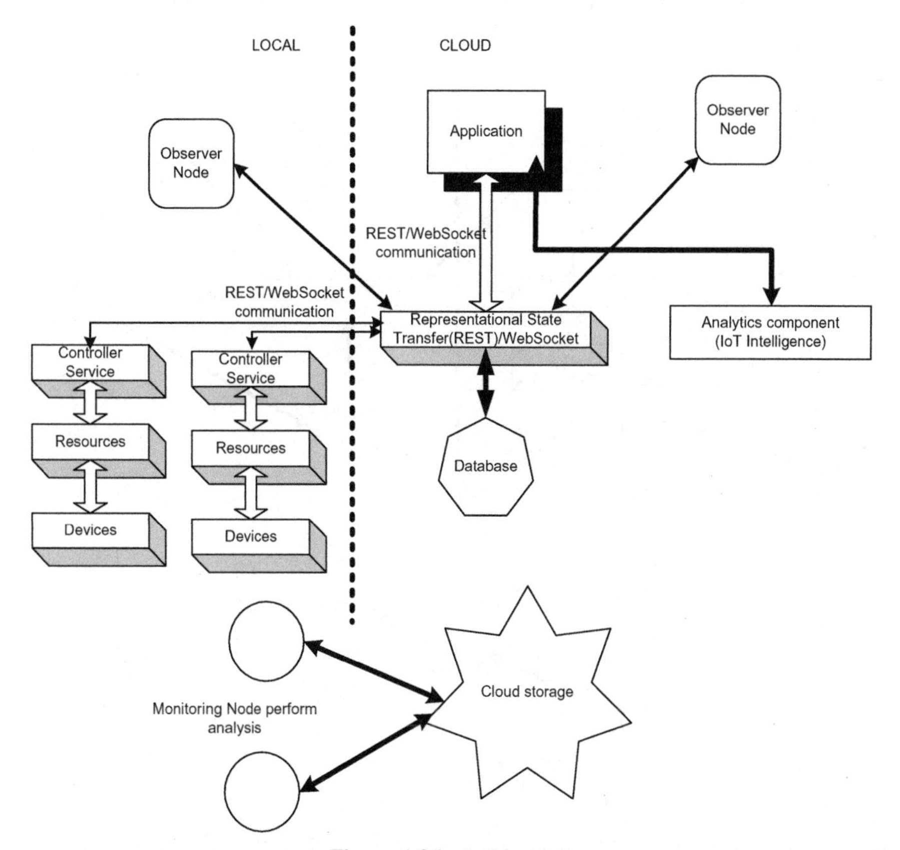

Figure 1.26 IoT level-4.

analyzed on the cloud. The system is suitable for WSN-based solutions. The best example is forest fire detection. The system comprises various nodes placed at different locations of the forest to monitor the fire, temperature, humidity, and CO2 levels.

1.10.6 Level-6 IoT System

A level-6 IoT system has multiple independent end nodes that perform sensing and/or actuation and send data to the cloud (Figure 1.28). Data are the stored on the cloud. The analytics components analyze the data and store the results in the cloud database. The results are visualized with the cloud-based application. The centralized controller is aware of the status of all the end nodes and sends control commands to the nodes. The example is weather

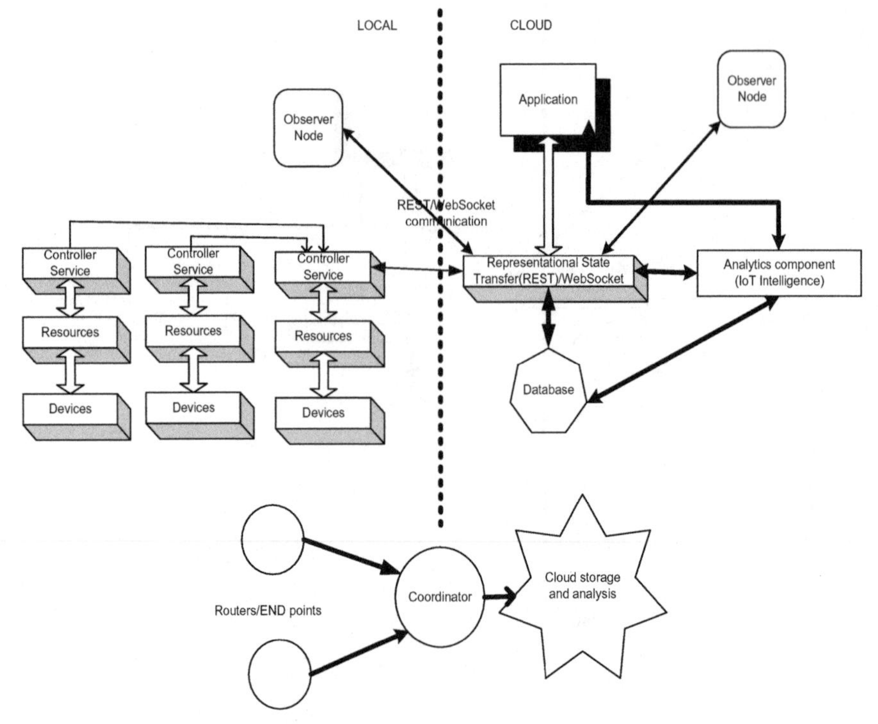

Figure 1.27 IoT level-5.

monitoring system which consists of multiple nodes placed at the different locations for monitoring temperature, humidity pressure, radiation, and wind speed. The end nodes are equipped with various sensors. The end nodes send the data to the cloud in real time using WebSocket service. The data are stored in the cloud-based server [12].

1.11 Domain-Specific IoT and Applications

Figure 1.29 shows the application domains of IoT cloud platforms in various fields. The major application domains of IoT are deployment management, monitoring management, visualization, research, application domain, device management, system management, heterogeneity management, data management, analytics, etc [8].

Figure 1.30 shows the broad application domains of IoT cloud platforms in various fields. The various fields are RFID, SoA, WSN, and supply chain

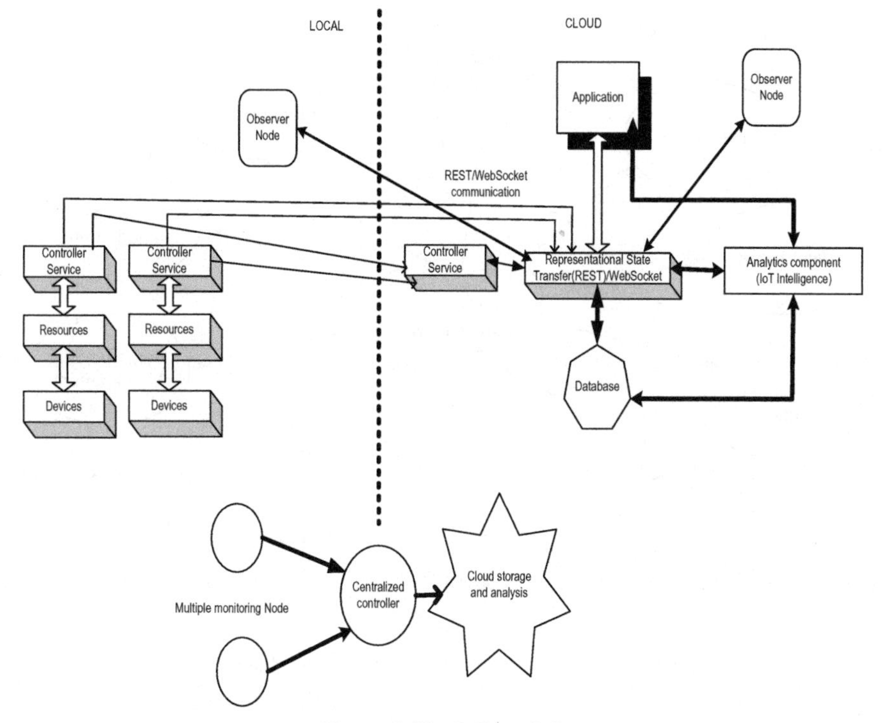

Figure 1.28 IoT level-6.

management (SCM), healthcare, smart society, cloud service, social compute, and security. Further all fields also subdivided into different areas like WSN category applications are environment, agriculture, infrastructure, etc [6].

1.11.1 IoT Application in Transport/Logistics

In transportation logistics, supply chain plays a major role while delivering goods from the origin to the destination. In order to control the movement and ensure supply chain, transparent IoT helps through global positioning and automatic identification of freight. IoT brings paradigm shift in SCM globally through intelligent freight movement.

1.11.2 IoT Application in the Smart Home

The major aspects for smart homes can be taken care like resource utilization (water, energy etc.), security, and comfort. To achieve comfort, these smart

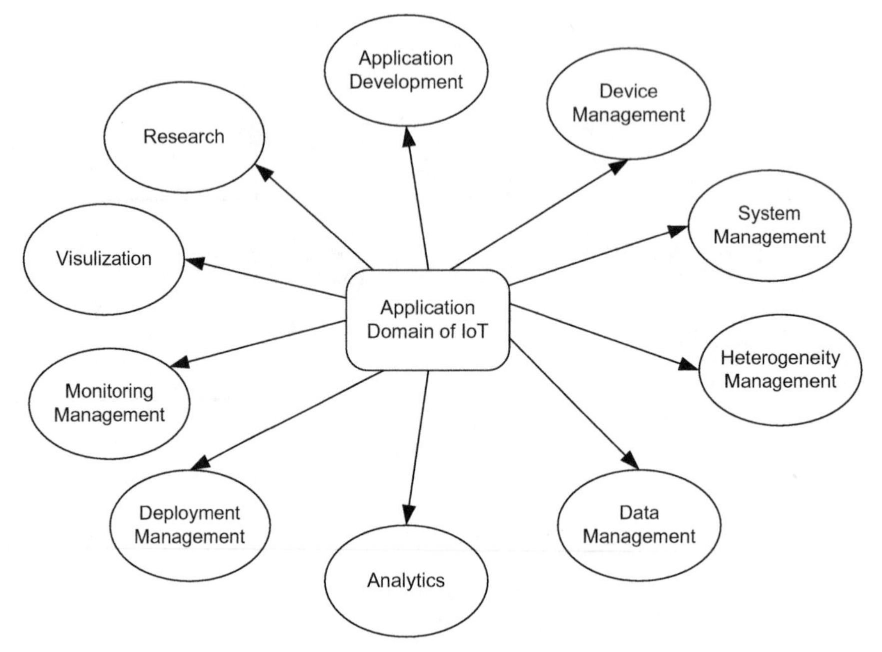

Figure 1.29 Application domains of IoT cloud platforms.

homes will reduce the overall expenditure by eliminating waste. In order to identify unauthorized entries, complex security system is required.

Figure 1.31(a) shows the Nest Learning Thermostat. This thermostat is making use of IoT concepts and able to reduce 15% on cooling bills and 12% on heating bills as an average. Figure 1.31(b) shows the Philips Hue as other example of smart homes. In this system, the bulb changes 600–800 color lumens as per mood of the occupant in that room.

Figure 1.31(c) shows the air quality sensing network made up by using a DIY sensor. The system senses the CO and NO_2 and other pollutants in home environment and determines the air quality. Figure 1.31(d) shows other home appliances like the Amazon Echo. The seven microphones are inbuilt and being highly sensitive, listener of words from across various noises, and answer the same [8].

1.11.3 IoT Application in Smart Cities

The smart cities mean smart waste and recycling process, traffic congestion, wireless outdoor lighting system, and smart parking system. Some of the

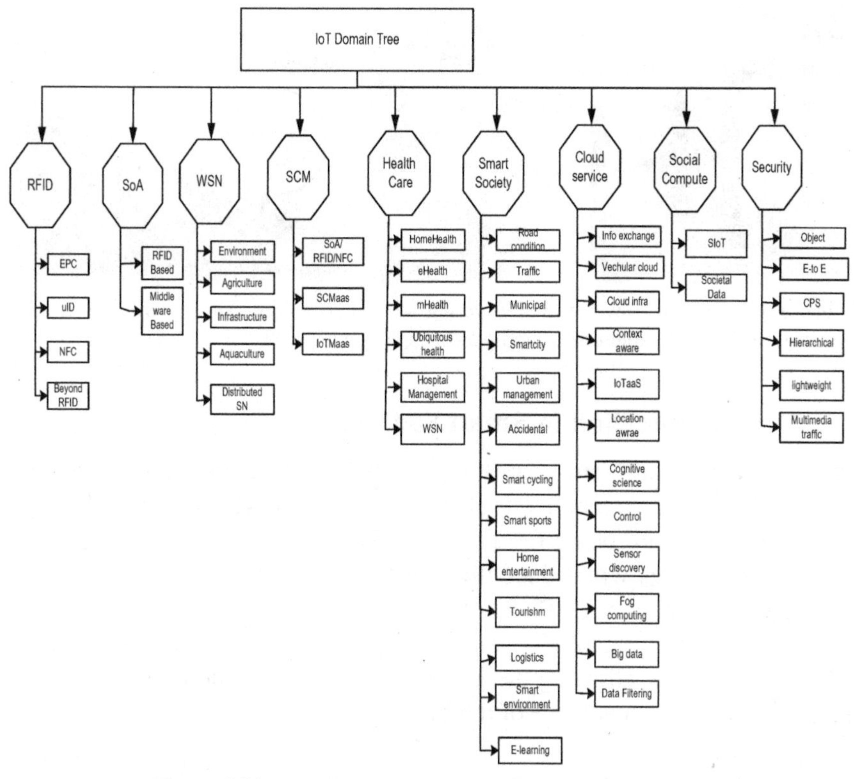

Figure 1.30 Application domains of IoT cloud platforms.

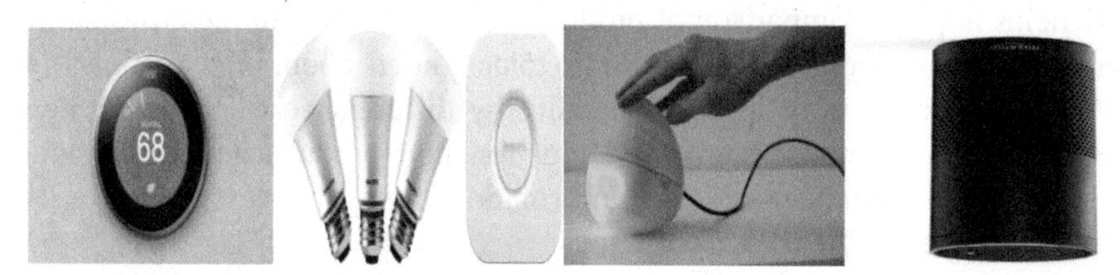

Figure 1.31 (a) Nest learning Thermostat. (b) Philips Hue. (c) Air quality sensing network. (d) Amazon Echo.

characteristics of smart cities are like economy, people, governance, mobility, environment, and living. Figures 1.32(a)–(c) show the Big belly smart waste and recycling system, city sense-wireless outdoor lightening system, and Libelium's smart parking solution, respectively [7].

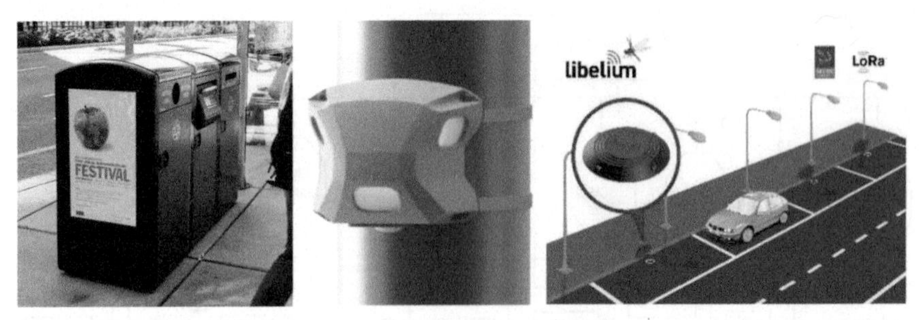

Figure 1.32 (a) Smart waste and recycling system. (b) City sense – wireless outdoor lightening system. (c) Libelium's smart parking solution [7].

1.11.4 IoT Application in Smart Factory

Smart factory means smart machinery maintenance, operating expenses (OPEX), ERP, and attendance logger. In the present scenario global supply chain, RFID tags are being used to track the products. Due to these consequences, companies will reduce their OPEX and improve their productivity by proper integration of ERP and other allied systems.

Maintenance of machinery can be facilitated by IoT technology through different sensors by allowing the capturing of real-time monitoring of health and performance of machines of the factory.

1.11.5 IoT Application in Retail

IoT helps in price comparison of products in order to benefit the customer as well as business people. It provides info to customers about price comparison of different products of the same quality by different vendors, which help the customers to choose the best vendor for their product, and at the same time, enterprise will have real-time information about what customer expects. Figure 1.33 shows the device for retails.

1.11.6 IoT Application in E-Health

Control and prevention are the two main objectives being used since long time and for long time in healthcare area. IoT enables the doctors to monitor their patients health from outstation. IoT enables more interactions in an efficient manner between the patients and doctors, and not only limited to localization but also through globalization. Important stakeholders of IoT-enabled health services are private and public hospitals. Figure 1.34 shows the smart health system.

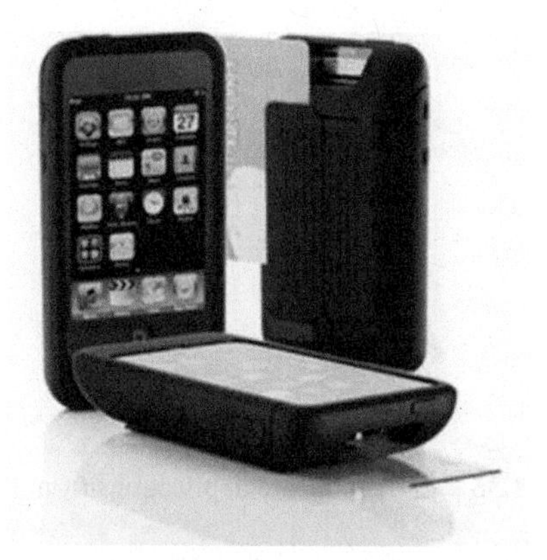

Figure 1.33 Device for retails.

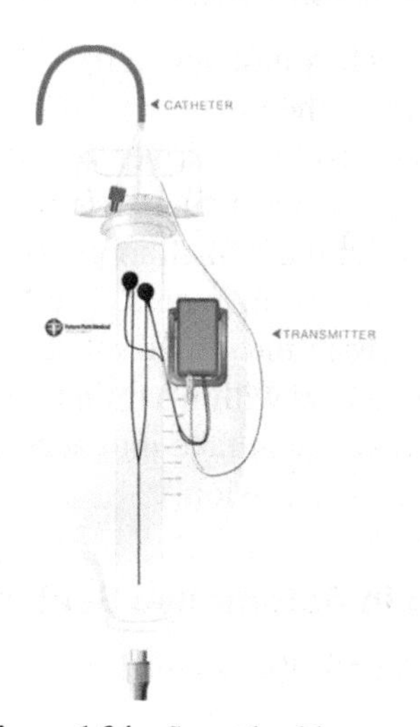

Figure 1.34 Smart health system.

Philips is one of those tech giants which are making full use of IoT opportunities available for business. Medication Dispensing Service is one of the most successful IoT healthcare applications from Philips. Focused around elderly patients who find it difficult to maintain their medication dosage on their own, MDS dispenses pre-filled cups as per the scheduled dosage. It notifies automatically when it's time to take medicine, refill, and malfunctioning or misses dosage. Figure 1.35 shows the IoT healthcare applications from Philips.

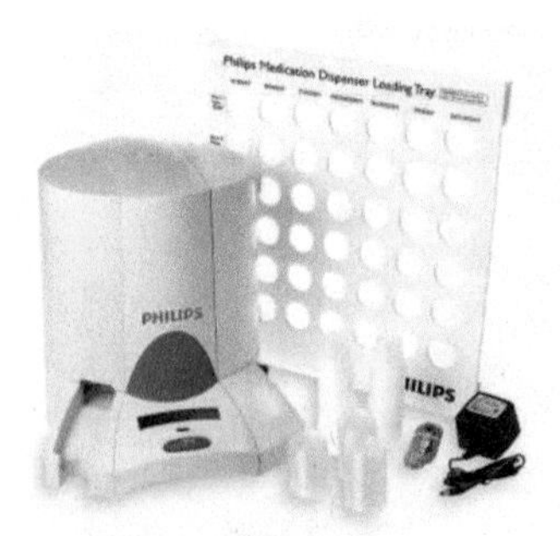

Figure 1.35 IoT healthcare applications from Philips.

1.11.7 IoT Application in Railroads

A train can be so long that its locomotives start to climb one hill while its mile of coal cars are still descending the last one. Cruise control that anticipates the terrain can save lots of fuel (just like a driver who practices "hypermiling"23 to save gas). It can save even more fuel if it knows something about the urgency of a train's schedule and the likelihood that the train will need to pull onto a siding to wait for another train to pass. The industrial internet promises to encompass entire railroads in integrated models that optimize everything from the placement of freight cars within a train to small variations in throttle. Delivered as a service, software can take into account an enormous range of contextual data to inform every decision.

1.11.8 IoT Application in Automotive Sector

The following are the IoT applications developed in automotive sector.

Google Car – Google started working on autonomous cars; they started this project with 10 numbers of cars. In April 2014, they announced that all 10 vehicles completed 1,126,541 km without any major incidents.

Kiva systems – Highly automated self-optimizing warehouse robots are successfully running different fulfillment centers across the world. Amazon acquired both kiva and zappos system at $ 0.75 and $ 1.2 billion, respectively.

Automated Guided vehicles – These vehicles are capable of having a payload of 70 tones and can be controlled precisely by using management and navigation software and transponders in the terminal road surface with a positioning accuracy of ±25 mm accuracy [11].

1.11.9 IoT Application in Manufacturing

Manufacturing is becoming broadly accessible to innovators operating at small scale. Sophisticated prototyping facilities are available at minimal cost in maker spaces across the world, where anyone with a modestly technical mindset can make use of newly simple tools – not only microcontrollers like Arduino, but also 3D printers, laser cutters, and CNC machine tools. Powerful computer hardware – controllers, radios, and so forth – has become so inexpensive that, at least at the outset, nearly any problem can be reduced to a control challenge that can be solved with software. Large-scale manufacturing will be beneficial from similar trends that will make it ever easier to bring intelligence to big machines. Intelligent software will make manufacturing more accurate and more flexible. Processors that are powerful enough to handle real-time streams of sensor data and apply machine-learning algorithms are now cheap enough to be deployed widely on factory floors to support such functions as machine-wear detection and nuanced quality-control observation. Logistics tools that transmit real-time data on shipments and inventory between manufacturers, shippers, and customers will continue to reduce inventory costs.

1.11.10 IoT Application in Wearables

Wearables are one of the hottest trends in IoT currently. Apple, Samsung, Jawbone, and plenty of others all are surviving in a cut throat competition. Wearable IoT tech is a very large domain and consists of an array of devices. These devices broadly cover the fitness, health, and entertainment requirements. The prerequisite from IoT technology for wearable applications is to be highly energy efficient or ultralow power and small sized. Here are some top examples of wearable IoT devices that fulfill these requirements.

Figure 1.36(a) shows the health tracker band and an excellent IoT application example in healthcare as well as wearable. It comes with features like

Figure 1.36 (a) Health tracker band. (b) Charge HR. (c) Motorola Moto 360 Sport.

activity tracking, food logging, and sleep patterns. In addition, it is offered in many styles and colors. It has features like activity tracking, sleep tracking, and smart coach. Figure 1.36(b) shows the charge HR and a high-performance IoT wearable which is provided with many smart features. It tracks heart rate as well as activities sitting on wrist. It provides the capability to automatically track heart rate, track workouts, monitor-sleeping pattern, get call notifications, and synchronize data with your PC and hundreds of Smart Phones wireless and many more. Figure 1.36(c) shows the Motorola Moto 360 Sport. It is time to get healthy space personalized even without smart phone. Motorola Moto 360 Sport is designed with this fact in mind. It delivers all the important information that is required from phone directly. It supports both Android and iOS apps.

1.11.11 IoT Application in Agriculture

Agriculture sector needs very institutive as well as highly scalable technology solutions. IoT applications can deliver the same to farmers. The agriculture sector applications are wine quality enhancing, green houses, golf courses meteorological station network, compost, etc.

1.11.12 IoT Application in Energy Management

Power grids of the future will not only be smart enough but also highly reliable. Smart grid concept is becoming very popular. The basic idea behind the smart grids is to collect data in an automated fashion and analyze the behavior or electricity consumers and suppliers for improving efficiency as well as economics of electricity use.

Figure 1.37 shows the Landis+Gyr Home energy management. The advanced metering will make energy management easier for everyone. Landis+Gyr are a wide range of energy management products. The smart metering solution offered by Landis+Gyr enables consumers, to better understand their energy needs as well help them with load management. They have many multi-energy metering solutions to offer for reliable and efficient energy management.

Figure 1.38 shows the Landis+Gyr grid management and the solutions are smart programs which provide capabilities to automate, analyze, as well as response to energy requirements in a smarter manner. They offer leading-edge tools that help both suppliers and consumers to reduce peak use problem and increase energy use efficiency.

Figure 1.37 Landis+Gyr Home energy management.

Figure 1.38 Landis+Gyr Grid management.

1.11.13 IoT Application in Industrial Automation

Industrial automation is one of the most profound applications of IoT. With the help of IoT infrastructure backed with advanced sensor networks, wireless connectivity, innovative hardware, and machine-to-machine communication, conventional automation process of industries will transform completely. IoT automation solutions for industries from all big names like NEC, Siemens, Emerson, and Honeywell are already in the market. The major area of work includes Smart structure, machine auto-diagnosis and assets control, indoor air quality, temperature monitoring, ozone presence, indoor location, and vehicle auto-diagnosis.

1.11.14 IoT Application in Smart Grids

Smart grid is a special application of IoT, future smart grid promises to use information about the behaviors of electricity suppliers and consumers in an automated fashion to improve the efficiency, reliability, and economics of electricity. Over 41,000 monthly Google searches highlight the concept's popularity.

1.11.15 IoT Application in Smart Supply Chain

Supply chains have been getting smarter for some years already. Solution for tracking goods, while they are on the road, or getting suppliers to exchange inventory information has been on the market for years. So while it is perfectly logic that the topic will get a new push with the IoT, it seems that so far its popularity remains limited.

1.11.16 IoT Application in Smart Farming

Smart farming is an often-overlooked business case for the IoT because it does not really fit into the well-known categories such as health, mobility, or industrial. However, due to the remoteness of farming operations and the large number of livestock that could be monitored, the IoT could revolutionize the way farmers work. However, this idea has not yet reached large-scale attention. Smart farming will become the important application field in the predominantly agricultural-product exporting countries.

1.11.17 IoT Application in Industrial Internet

The industrial internet is also one of the special IoT applications. While many market researches such as Gartner or Cisco see the industrial internet as the IoT concept with the highest overall potential, its popularity currently does not reach the masses like smart home or wearables do. The industrial internet, however, has a lot going for it. The industrial internet gets the biggest push of people on Twitter ($\sim$1,700 tweets per month) compared to other non-consumer-oriented IoT concepts.

1.11.18 IoT Application in Connected Car

The connected car is coming up slowly. Owing to the fact that the development cycles in the automotive industry typically take two to four years, we have not seen much buzz around the connected car yet. Most large automakers as well as some brave startups are working on connected car solutions. In addition, if the BMWs and Fords of this world do not present the next-generation Internet-connected car soon, other well-known giants will: Google, Microsoft, and Apple have all announced connected car platforms.

1.11.19 IoT Application in Connected Health

The connected health means digital-health, telehealth, telemedicine, etc. These are the major areas of work. The concept of a connected healthcare system and smart medical devices bears enormous potential not just for companies also for the well-being of people in general.

1.11.20 IoT Application in Poultry

The major applications are livestock monitoring, cattle health monitoring, and tracking. Using IoT applications to gather data about the health and well-being of the cattle, ranchers knowing early about the sick animal can pull out and help prevent a large number of sick cattle.

1.11.21 IoT Application in Smart Environment

The areas of research are forest fire detection, air pollution, snow level monitoring, landslide and avalanche prevention, and earthquake early detection.

1.11.22 IoT Application in Security and Emergency

The areas of research are perimeter control, radiation presence, explosive, and hazardous gases.

1.11.23 IoT Application in Smart Animal Farming

The areas of research include hydroponics, offspring care, animal tracking, toxic gas levels, etc.

1.11.24 IoT Application in Smart Water

The areas of research include portable water monitoring, chemical leakage detection, swimming pool remote measurement, pollution levels, water leakages, and river floods.

1.12 IoT Servers

IoT is merging of various "things" with the use of Internet to establish a smart connection between people and smart objects. Cloud is one of the major components of IoT, which provides application-specific services in many domains. Currently, many cloud providers are in the market, which provides suitable IoT-based services for the specific applications.

1.12.1 KAA

KAA is an open source middleware IoT platform with Apache License 2.0 for building smart connections for end-to-end IoT solutions. It provides services for data exchange between the connected devices, data analytics, visualization, and IoT cloud services. It supports *NoSQL and* Big Data base applications supported, but the major disadvantage is its less support to hardware modules [http://www.kaaproject.org/].

1.12.2 Carriots

Carriots is platform, helping anyone to build quick IoT applications. It saves time, cost, and troubles. Platform as a Service (PasS) cloud model is featured with services like remote device management and control, rule-based listeners' activity logging, triggering custom alarms, and data export. It main advantage is its usability in triggering-based applications, but it is less user friendly [https://carriots.com].

1.12.3 Temboo

Temboo is a cloud-based platform for application code generation. It involves less wiring and coding of hardware and software which results in development of products in less time. It has more than 90 inbuilt libraries named "Choreos" for third-party services including Yahoo weather, Amazon cloud, Twitter micro blogging, Twilio telephony, Ebay product shopping, Flickr photo management, Facebook Graph API, Google analytics, PayPal payment, Uber vehicle confirmation, YouTube video streaming, and many more. It supports all chores-based applications but not suitable for resources-intensive applications [https://temboo.com].

1.12.4 SeeControl IoT

SeeControl is an IoT cloud platform which is specialized at device messaging and management. Sensor data visualization, analytics, and complete work flow monitoring can be done by SeeControl. It has open API-based push/pull architecture for scalable IoT products. Its major advantage is its support to push/pull devices, but its visualization is not too good [http://www.seecontrol.com].

1.12.5 SensorCloud

SensorCloud is an IoT cloud which provides PasS to acquire, visualize, monitor, and analyze the data received from wired or wireless sensors. SensorCloud is a powerful tool for cloud computing facilities like data scalability, rapid visualization, and user program analysis. It allows developers to perform complex mathematical operations on the data. It can manage a large pool of sensor devices but do not serve in open source devices [http://www.sensorcloud.com].

1.12.6 Etherios

Etherios supports comprehensive products and services for the connected enterprises. Its cloud is designed on the PaaS model to enable users for connecting product and gain real-time visibility into their assets. Etherios provides the connectivity for modern enterprises and facilitating through thousands of off-the-shelf wired and wireless solutions designed for a specific purpose. It is specialized cloud, but the developers are restricted with limited devices [http://www.etherios.com].

1.12.7 Xively

Xively is a Gravity Cloud technology-based enterprise IoT cloud service. This LogMeIn owned platform helps companies to manage their product business by addressing a number of practical needs by scalable, secure, and reliable connectivity. It also features the right business data processing services to its IoT-enabled customers through flexible API connectors. It is easy to integrate with devices but has minimum notification services [https://xively.com].

1.12.8 Ayla's IoT Cloud Fabric

Ayla IoT fabric is a PaaS modeled enterprise class. It is a simple and cost-effective solution for OEMs for connecting any device to the Internet. Ayla Networks provide software agents embedded in both devices and mobile device applications for end-to-end support. Ayla's Agile Mobile Application Platform is built with its mobile libraries that provide an optimized mobile APP for iOS and Android users. It provides easy mobile application development platform but not suitable for small-scale developers [https://www.aylanetworks.com].

1.12.9 thethings.io

thethings.io is a platform which provides a complete back-end solution for IoT APP developers through an easy and flexible API. thethings.io is hardware agnostic which allows to connect any device that is capable of using HTTP, Websockets, MQTT, or CoAP protocols. Real-time, rule-based jobs can be easily monitored by developing end-to-end connectivity, but it lacks in self-sustenance [https://thethings.io].

1.12.10 Exosite

Exosite is modular, enterprise-grade IoT software platform which helps to bring connected products in market. It has cloud platform based on IoT Software as a Service (SaaS) which provides real-time data visualization and analytics support to the users. It is a hosted server-based system which has web service enabled APIs, built in infrastructural framework, lightweight, and flexible back-end conjugated with UDP, HTTP, and JSON RPC. The system development is easy with it but it lacks in big data provisions [https://exosite.com].

1.12.11 Arrayent Connect TM

Arrayent is an IoT platform which enables heterogeneous brands like Whirlpool, Maytag, and First Alert to connect users' products to smart handheld devices and web applications. Arrayent Connect Cloud is an IoT operating system which is based on the SaaS model. It helps hosting all devices with Over-the Air firmware updates in low data latency rate. Further, its secure, reliable, and scalable data sources help users to get retrieved, processed, and delivered. It is flexible in use but it lacks in triggering-based services [http://www.arrayent.com].

1.12.12 OpenRemote

OpenRemote is an open source IoT middleware solution which allows users to integrate any device – protocol – design using available resources like iOS, Android, or web browsers. A user can design tools for developing completely customized solutions by using OpenRemote's cloud service, which leverages to integrate a variety of protocols from WiFi to ZigBee. It supports to open cloud services but has high cost [http://www.openremote.com].

1.12.13 Arkessa

Arkessa provides services like overall connectivity, monitoring, control, and management between IoT-based devices and enterprises. Arkessa's mission is to empower companies to involve into the IoT for the development of new revenue streams through improved customer satisfaction. It provides enhanced potential values to received data streams from remote devices. Arkessa follows the PaaS model to formulate a single enterprise management portal and integrates machine data streams with the available CRM, ERP, big data, and other analytics systems for efficient and optimized device management services. It has enterprise enabled design facet but its visualization apps are not proper [http://www.arkessa.com].

1.12.14 Oracle IoT Cloud

It comprises four crucial parameters. It performs operations on received data including analysis, acquisition, and integration. It supports database but lacks in connectivity of open source devices [https://cloud.oracle.com/iot].

1.12.15 Nimbits

Nimbits is a cloud server which provides solutions to Edge Computing IoT-related services. It performs operations like noise filtering and sends data on the cloud. It is easy to adopt but lacking in the real-time processing of query [http://www.nimbits.com].

1.12.16 ThingWorx

ThingWorx is a data-driven decision-making cloud. It provides M2M and IoT services based on SQUEAL. Zero coding facility is available. It has easier data-intensive application building but a number of devices are limited [https://thingworx.com].

1.12.17 InfoBright

InfoBright is an IoT-based analytical database platform, which connects business to store and act on machine-generated data for a complete eco system [https://www.infobright.com/index.php/internet-of-things].

1.12.18 Jasper Control Center

Jasper Control Center is a platform based on Japer control. Control center is designed to automate the connected devices and help to analyze real-time behavior patterns. The main target of this is manufacturing, security, transportation, and home automation. The main advantage is its rule-based behavior pattern [https://www.jasper.com].

1.12.19 Echelon

Echelon is an IIoT-based platform for cloud with resources like microphones, hardware devices, and other applications. It addresses the fundamental requirements of IIoT. It is good for industrial prospective but lacks in basics for the beginners [http://www.iiot.echelon.com].

1.12.20 AerCloud

AerCloud platform collects, manages, and analyzes sensory data for IoT and M2M applications. It ensures security and reliability by enabling the applications through seamless scalability. It is also scalable to M2M services but not suitable for the developers [http://www.aeris.com].

1.12.21 ThingSpeak

ThingSpeak is an open IoT data platform which is based on public cloud technology. It has open API which enables receiving of real-time data. It has data storage, monitoring, and visualization facilities. It has been developed by Mathworks. It has triggering facility with public cloud, but has limitation of devices which can be connected simultaneously [https://thingspeak.com].

1.12.22 Plotly

Plotly is a data visualization cloud service provider for public. It provides data storage, analysis, and visualization services. Python, R, MATLAB, and Julia-based APIs are implemented in Plotly. It is a good visualization tool for IoT but with a limited storage facility [https://plot.ly].

1.12.23 GroveStreams

GroveStreams is a public cloud for data visualization. It supports various data types. It enables seamless monitoring but lacks in statistical services [https://thingworx.com].

1.12.24 Microsoft Research Lab of Things

Lab of Things is a platform developed by Microsoft. It was developed for experimental research for academic institutions. It is used for making connections between devices for applications like home automation, energy, healthcare, etc., but it lacks in IoT-supported API [http://www.lab-of-things.com].

1.12.25 IBM IoT

IBM IoT is an organized architecture cloud platform. It supports complex industry solutions. It can enable device identity but application prototyping is difficult [https://internetofthings.ibmcloud.com].

1.12.26 Blynk

It is a platform with iOS and Android apps to control Arduino and Raspberry Pi over the Internet. It supports graphical interface to build projects just by dragging the widgets. It supports many IoT modules and ready for IoT.

1.12.27 Cayenne APP

Cayenne is an App for smartphones and computers which allows controlling the Raspberry Pi and Arduino through the use of a graphical interface. It can add and control sensors, motors, actuators, GPIO boards, and more. It has customizable dashboards with drag-and-drop widgets for connection devices. It supports quick and easy setup.

1.12.28 Virtuino APP

Virtuino platform creates amazing virtual screens on smart phones or tablets to control the automation system created with arduino or similar boards. It supports Arduino and can be connected with the HC-05 Bluetooth, Ethernet Shield, and ESP8266 modules. It supports monitoring sensors values from the IoT server ThingSpeak.

1.13 Internet of Things Device Design Methodology

The steps to design the IoT-related system are follows.

Step 1: Purpose and requirements specification: The first step in IoT system design methodology is to define the purpose and requirements of the system. In this step, the system purpose, behavior, and requirements (such as data collection requirements, data analysis requirements, system management requirements, data privacy and security requirements, user interface requirements, etc.) are captured.

This can be understood by applying this to, for example, a smart IoT-enabled robot system; the purpose and requirements for the system may be described as follows:

1. Purpose: The robot control system that allows controlling the direction of robots using IoT.
2. Behavior: The switches on APP provide the data which are 10, 20, 30.....serially. According to the received data, the command for robot to move forward, reverse, left, right and stop.
3. System management requirement: The system should provide remote monitoring and control functions.
4. Data analysis requirement: The system should perform local analysis of the data.
5. Application deployment requirement: The application should be deployed locally on the device, but should be accessible remotely.

Step 2: Process specification: The second step in the IoT design methodology is to define the process specification. In this step, the use cases of the IoT system are formally described based on and derived from the purpose and requirement specifications.

Step 3: Domain model specification: The third step in the design methodology is to define the domain model. The domain model describes the main concepts, entities, and objects in the domain of the IoT system to be designed.

Step 4: Information model specifications: The fourth step in the IoT design methodology is to define the information model. The information model defines the structure of all the information in the IoT system, for example, attributes of virtual entities, relations, etc. The information model does not describe the specifics of how the information is represented or stored.

Step 5: Service specifications: The fifth step in the design methodology is to define the service specifications. Service specifications define the services in the IoT system, service types, service inputs/output, service endpoints, service schedules, service preconditions, and service effects.

Step 6: The sixth step in the IoT methodology is to define the IoT level for the system.

Step 7: Functional group: The seventh step in the design methodology is to define the functional view. The functional view defines the functions of the

IoT systems grouped into various functional groups. Each functional group either provides functionalities for interacting with instances of the concepts defined in the domain or provides information related to these concepts.

Step 8: Operational view specifications: The eighth step in the IoT design methodology is to define the operational view specifications. In this step, various options pertaining to the IoT system deployment and operation are defined, such as service hosting options, storage options, Device options, application-hosting options, etc.

Examples for operational view specification for robot control using IoT system are as follows:

Devices: Computing device (NodeMCU), Motor driver L293D, and two DC geared motors.

Communication APIs: REST APIs.

Communication protocols: Link layer-802.11, Network layer-IPv4/IPv6, and Transport layer-TCP.

Application: HTTP.

Services:

Controller service: Hosted on device, implemented in Python, and run as a native service.

Mode Service: REST-ful web service, hosted on device, implemented with Django-REST framework.

State service: REST-ful web service, hosted on device, implemented with Django-REST framework.

1.14 Role of IoT in Automotive Industries

The IoT is breaking fresh ground for car manufacturers by introducing entirely new layers to the traditional concept of a car. This upgrade – the connected, smart car – comes as a revolutionary way for us to drive and stay in touch with the world around at the same time. By offering a fancy-free variety of infotainment services and connected car applications for drivers, the automotive industry has the potential to become an IoT champion among other industries. It may fuel the IoT cloud services' adoption among car owners and walkers alike. For companies in the automotive sector, entertainment, and maintenance service providers, Kaa offers a stack of plug-and-play IoT components that streamline the development of connected car applications by times and ensure smooth integration between separate modules of the connected car within a secure cloud environment. Kaa is highly scalable and can easily handle thousands of connected vehicles simultaneously, as well as automatically balance out peak loads in cloud service usage. With Kaa, it is easy to enable new services over the air and manage different service subscription plans and user groups. All of these technologies used within the car are interconnected and centrally controlled. Some of these, including car-to-infrastructure and positioning, are connectivity features in their own right. The car as a technology hub has already started fulfilling the Internet of Everything concept with people, processes, and things interacting seamlessly.

A complete picture can be obtained by collecting, managing, and analyzing data, and connecting everything to the Internet. The Internet of Cars turns into a complete platform in the Internet of Everything. Infotainment and telematics and safety and security are highly improved with the benefit of data transfer and connectivity. However, powertrain/fuel economy may not offer great benefits when working in a connected world.

From a connected vehicles' perspective, here are some of the major ideas which can be powered by IoT, and which may soon become very ordinary.

1. Crash Response: Connected cars can automatically send real-time data about a crash along with vehicle location to emergency teams. This can save lives by accelerating emergency response.
2. Car Problem Diagnosis: Connected cars are capable of generating prognostic data that can predict a problem before a part even fails, which would prevent the inconvenience of a breakdown and help consumers better manage the timing of vehicle care. Preventative maintenance promises to help reduce repair and warranty costs.
3. Convenience Services: The ability to access a car remotely makes possible services such as remote door unlock, find my vehicle, and stolen vehicle recovery.
4. Integrated Navigation: Connected cars can integrate GPS with online services to respond to driver preferences, routing, fuel availability and pricing, traffic alerts, points of interest, etc.
5. Traffic Management: Connected car technology can provide transportation agencies with improved real-time traffic, transit, and parking data, making it easier to manage transportation systems for reduced traffic and congestion.
6. Infotainment: Connected cars can provide online, in-vehicle entertainment options that provide streaming music and information through the

dashboard. AAA has called for limiting certain features while driving to prevent distractions.

7. Discounts and Promotional Offerings: Companies can provide insurance or location-based discounts and promotional offerings.
8. Enhanced Safety: Pilot programs for vehicle-to-vehicle (or "V2V") and vehicle-to-infrastructure ("V2I") communications are underway that will warn drivers of potential collisions, dangerous road conditions, and other impediments to safe travel. A range of crash prevention technologies integrated with connected communications such as intersection assistance likely will reduce the number of crashes in the coming years

1.15 Introduction to Arduino

Arduino is an open-source prototyping platform based on easy-to-use hardware and software. Arduino boards are able to read inputs like digital and analog sensors and turn it into an output like DC motor, solenoid, relay etc.

The advantages of Arduino platform are as follows:

Inexpensive – Arduino boards are relatively inexpensive compared to other microcontroller platforms.

Cross-platform – The Arduino Software (IDE) supports on Windows, Macintosh OSX, and Linux operating systems but most microcontroller systems are limited to Windows.

Simple, clear programming environment – The Arduino Software (IDE) is easy to use for beginners. It is also flexible enough for advanced users to developed complex application firmware.

Open source and extensible software – The Arduino software is an open source tool, can be expanded by adding C++ libraries with it, and adds AVR-C code directly into your Arduino programs if you want to.

Open source and extensible hardware – The Arduino boards are fall under a Creative Commons license. So own version can be made by adding more hardware with it.

Figure 1.39 Shows the view of Arduino nano board. The pins details are as follows.

Input and Output – Arduino unohas 14 digital pins that can be used as an input or output. These pins operate at 5 V and an individual provides or receives 40 mA current. It has an internal pull-up resistor of 20–50 kΩ. The Arduino Uno has six analog inputs named A0–A5. The resolution of ADC is

Figure 1.39 View of Arduino Nano.

10 bits (means 1,024 digital levels). In addition, there are some pins that have specialized functions as follows:

Serial: The Arduino uno has pins 0 (RX) for receive transistor–transistor logic (TTL) data and 1 (TX) for transmit TTL data using UART mode.

External Interrupts: The pins 2 and 3 are used as interrupt pins and can be used to read a rising or falling edge, or a change in value.

PWM: The pins 3, 5, 6, 9, 10, and 11 of Arduino Nano are used for pulsewidth modulation.

SPI: The 10 (SS), 11 (MOSI), 12 (MISO), and 13 (SCK) of Arduino uno are used as serial peripheral interface (SPI).

LED: Pin 13 of Arduino uno board has an inbuilt LED.

TWI: The pins A4 or SDA pin and A5 or SCL in Arduino nano are used as two-wire interface (TWI) or inter-IC communication (I2C).

AREF: Aref pin of Arduino nano board provides reference voltage for the analog inputs.

Reset: The reset pin is used to reset the microcontroller.

1.16 Introduction to NodeMCU

ESP8266 is a low-cost WiFi microchip with TCP/IP stack and microcontroller. ESP8266 was produced by Espressif Systems from the Chinese manufacturer from Shanghai. NodeMCU was created shortly after the ESP8266 came out. Its processor has an L106 32-bit RISC architecture running at 80 MHz. It is widely used in IoT applications. The NodeMCU-Amica is a C++-based firmware.

An Arduino core for the ESP8266 has been developed with WiFi SoC. This is popularly called as the "ESP8266 Core for the Arduino IDE" and it has become one of the leading software development platforms for various ESP8266-based modules and development boards, including NodeMCUs. Figure 1.40 shows the view of NodeMCU and Figure 1.41 shows the detailed pin description.

Figure 1.40 View of NodeMCU.

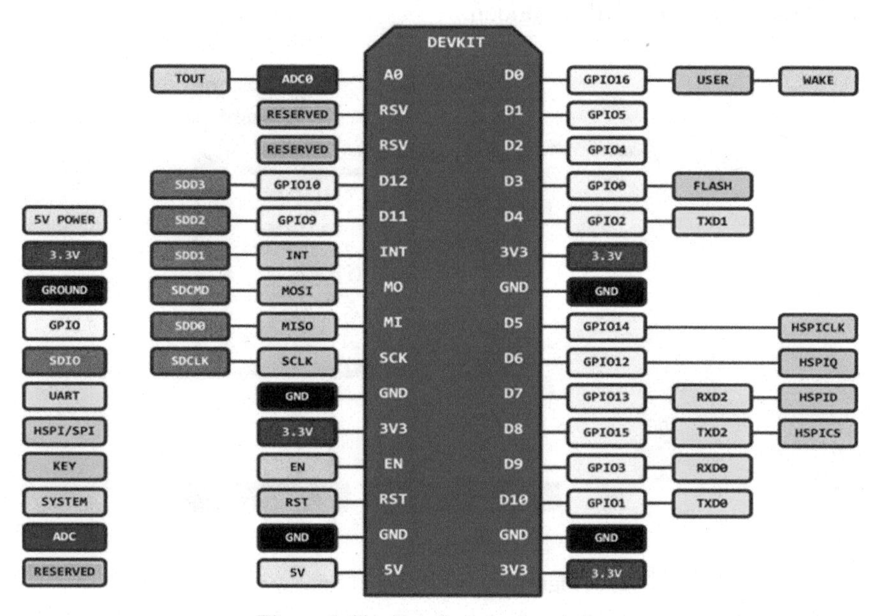

Figure 1.41 Detailed pin description.

Table 1.3 shows the GPIO (general purpose input/output) of NodeMCU.

Table 1.3 GPIO of NodeMCU

IO Index	ESP8266 Pins	IO Index	ESP8266 Pins
D0	GPIO16	D7	GPIO13
D1	GPIO5	D8	GPIO15
D2	GPIO4	D9	GPIO3
D3	GPIO0	D10	GPIO1
D4	GPIO2	D11	GPIO9
D5	GPIO14	D12	GPIO10
D6	GPIO12		

1.17 Introduction to GPRS

The GPRS/GSM Module MicroSIM card TTL Serial Port SIMCOM – HBK0004 [SIM800L] is designed for global market. It works on frequencies 850 MHz [GSM], 900 MHz [EGSM], 1,800 MHz [DCS], and 1,900 MHz [PCS]. SIM800 features GPRS multi-slot class 12/class 10 (optional) and supports the GPRS coding schemes CS-1, CS-2, CS-3, and CS-4. Figure 1.42 shows the view of the GPRS modem.

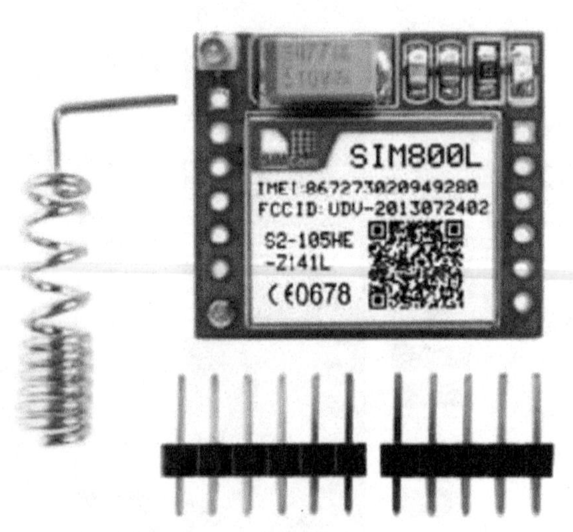

Figure 1.42 GPRS modem.

The features of the module are as follows:

1. Module Model: SIM800L Quad-band 850/900/1,800/1,900 MHz.
2. It can be interfaced with 8051/AVR/ARM/PIC/Arduino/Raspberry-pi.
3. It has GPRS multi-slot class 12 connectivity.
4. It is supported by AT Command.
5. It has a real-time clock on it.
6. Its supply voltage range is 3.4–4.4 V.
7. It supports 3.0–5.0 V logic level, which means low power consumption.
8. It has a current consumption of 1 mA in sleep mode.

The SendMessage() and ReadMessage() are two functions that are useful to send and receive messages, while it is connected with arduino. The SendMessage() is the function created in arduino IDE sketch to send an SMS. By sending "AT+CMGF=1" to the GPRS modem, it will come to text mode. For this, Serial.print() function is used. It writes data to the serial port. The number to which the message needs to be sent is set by the AT command "AT+CMGS=\"mobile no.\"\r." An SMS is sent in the next line. Each command follows a delay of 1 s.

AT commands to send an SMS are as follows:

1. Send AT+CMGF=1 using Serial.println command in Arduino IDE to set the GPRS/GSM module in text mode.
2. Send AT+CMGS=\"mobile no.\" \r using Serial.println command in Arduino IDE to send the message to assign a number.
3. Send (char)26; using Serial.println command in Arduino IDE which is ASCII of cntl+Z to stop the process.

The **RecieveMessage()** is the function to receive an SMS. The AT command to receive an SMS is "AT+CNMI=2,2,0,0,0" – just send this command to the GSM module and apply a 1 s delay. After this, send the SMS to the SIM card number inside the GSM module. To read the stored messages in the SIM, send the AT command – "AT+CMGL=\" ALL\"\r" to the module.

AT commands to receive an SMS using Arduino and GPRS/GSM module are given below:

1. Send AT+CMGF=1 command using Serial.println instruction in Arduino IDE to set the GSM module in text mode.
2. Send AT+CNMI=2,2,0,0,0 command using Serial.println instruction in Arduino IDE to receive the SMS.

2

Interfacing of Arduino with Input/Output Devices

This chapter describes the interfacing of Arduino with input/output devices like digital sensors, analog sensors, and serial communication.

2.1 Digital Sensor – Capacitive Touch Proximity Sensor

This section shows the interfacing of Arduino NANO with a digital sensor.

2.1.1 Introduction

Figure 2.1 shows the block diagram of the Arduino NANO and external devices like liquid crystal display (LCD) and digital sensors. It comprises a +12 V/500 mA power supply, an Arduino Nano, an LCD, and a capacitive touch sensor (digital sensor). The objective of the system is to display the sensory data on the LCD by reading the capacitive touch sensor and make LED ON/OFF.

Figure 2.2 shows the view of the capacitive touch sensor [Sunrom part −4441]. The output can be configured as active high or low as per the requirement. The operating voltage is 2–5.5 V. The calibration delaytime is 0.5 s. Table 2.1 shows the component list required to develop the system. Figure 2.3 shows the circuit diagram of the system.

2.1.2 Circuit Diagram

The following are the interfacing connections of NodeMCU and the I/O devices:

1. +5V and GND pins of the Arduino Nano are connected to +5V and GND pins of the power supply.
2. Pins 1 and 16 of the LCD are connected to GND of the power supply.

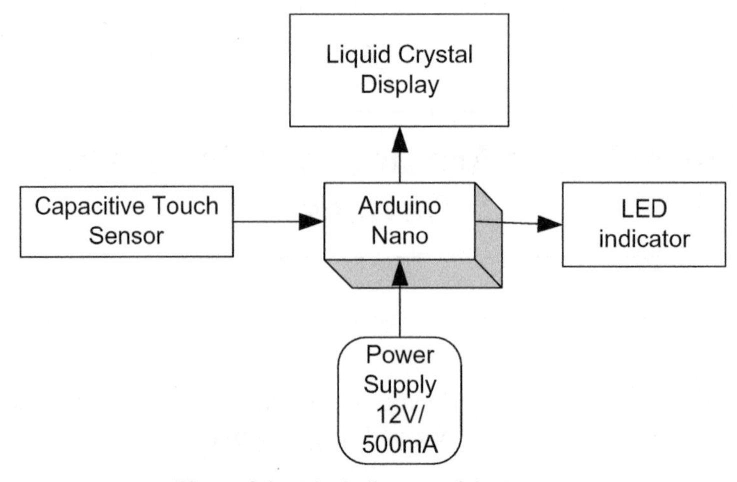

Figure 2.1 Block diagram of the system.

Figure 2.2 Capacitive touch sensor.

Table 2.1 Component list

Component	Quantity
Power supply 12 V/1 A	1
Arduino Nano	1
Jumper wire M-M	20
Jumper wire M-F	20
Jumper wire F-F	20
Power supply extension (to get more +5V and GND)	1
Level converter to 12 to 5 V, 3.3V	1
Capacitive touch sensor	1
+12V to +5V convertor	1
LCD20*4	1
LCD breakout board/patch	1

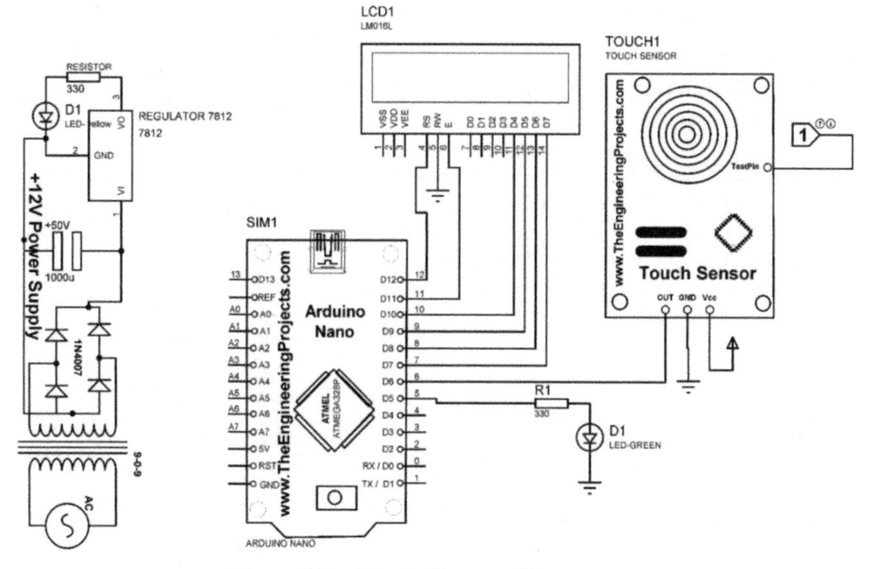

Figure 2.3 Circuit diagram of the system.

3. Pins 2 and 15 of the LCD are connected to +5V of the power supply.
4. Fixed terminals of the 10K POT are connected to +5V and GND of the power supply and variable terminal to pin 3 of the LCD.
5. Pin 12, GND, and pin 11 of the Arduino Nano are connected to pin 4(RS), pin 5(RW), and pin 6(E) of the LCD, respectively.
6. Pin 10, pin 9, pin 8, and pin 7 of the Arduino Nano are connected to pin 11(D4), pin 12(D5), pin 13(D6), and pin 14(D7) of the LCD, respectively.
7. +Vcc, GND, and OUT pins of the touch sensor are connected to +5V, GND, and pin 6 of the Arduino Nano, respectively.

2.1.3 Program Code

```
/////// Library for LCD16*2
#include <LiquidCrystal.h>
LiquidCrystal lcd(12, 11, 10, 9, 8, 7);
/////// LED and Sensor Pins
const int Touch_Sensor_Pin = 6;
const int LED_pin =  5;
int Touch_sensor_logic = 0;
void setup()
```

```
{
 lcd.begin(16, 2);// Initialize LCD 16*2
 pinMode(LED_pin, OUTPUT);// set pin 5 as output
 pinMode(Touch_Sensor_Pin, INPUT_PULLUP);// set pin 6 as
    input when sensor is in active LOW
 lcd.setCursor(0,0);// set cursor on LCD
 lcd.print("Touch sensor");// print string on LCD
 lcd.setCursor(0,1);// setcursor on LCD
 lcd.print("based system");// print string on LCD
 delay(2000);// provide delay of 2Sec
}

void loop()
{
   Touch_sensor_logic= digitalRead(Touch_Sensor_Pin);
   if (Touch_sensor_logic == LOW)
   {
     lcd.clear();
     lcd.setCursor(0,1);
     lcd.print("Touch Detected");
     digitalWrite(LED_pin, HIGH);
     delay(20);
   }
   else
   {
     lcd.clear();// clear LCD
     lcd.setCursor(0,1);// set the cursor on LCD
     lcd.print("No Touch");// Print string on LCD
     digitalWrite(LED_pin, LOW);// make pin as LOW
     delay(20);// provide delay of 20msec
   }
}
```

2.2 Analog Sensor – DC Voltage Sensor

This section describes the interfacing of the analog sensor with the Arduino NANO.

2.2.1 Introduction

Figure 2.4 shows the block diagram of the Arduino NANO and external devices like LCD and analog sensors. It comprises a +12 V/500 mA power supply, an Arduino Nano, an LCD, and a DC voltage sensor. The main objective is to measure and display the value of the DC voltage sensor on the LCD.

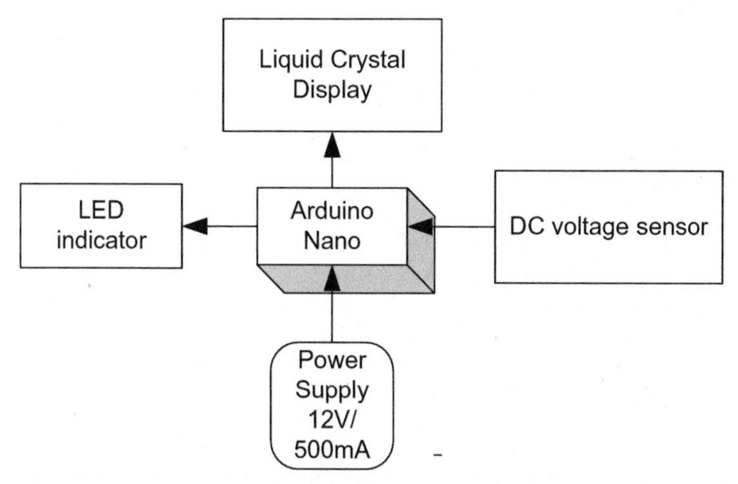

Figure 2.4 Block diagram of the system.

Figure 2.5 Voltage sensor.

The voltage sensor is connected through the voltage divider circuit. This sensor is capable of reading upto 25 V DC voltage from the source. The analog voltage resolution is 0.00489 V. The voltage detection range is 0.02445–25 V. Figure 2.5 shows the view of the voltage sensor. Table 2.2 shows the component list to develop the system.

2.2.2 Circuit Diagram

The connections of the system are as follows:

1. +5V and GND pins of the Arduino Nano are connected to +5V and GND pins of the power supply.
2. Pins 1 and 16 of the LCD are connected to GND of the power supply.

Table 2.2 Component list

Component	Quantity
Power supply 12 V/1 A	1
Arduino Nano	1
Jumper wire M-M	20
Jumper wire M-F	20
Jumper wire F-F	20
Power supply extension (to get more +5V and GND)	1
Level converter to 12 to 5 V, 3.3 V	1
DC voltage sensor	1
+12V to +5V convertor	1
LCD20*4	1
LCD breakout board/patch	1

3. Pins 2 and 15 of the LCD are connected to +5V of the power supply.
4. Fixed terminals of the 10K POT are connected to +5V and GND of the power supply and variable terminal to pin 3 of the LCD, respectively.
5. Pin 12, GND, and pin 11 of the Arduino Nano are connected to pin 4(RS), pin 5(RW), and pin 6(E) of the LCD, respectively.
6. Pin 10, pin 9, pin 8, and pin 7 of the Arduino Nano are connected to pin 11 (D4), pin 12(D5), pin 13 (D6), and pin 14 (D7) of the LCD, respectively.
7. +Vcc, GND, and OUT pins of the DC voltage sensor are connected to +5V, GND, and A0 pin of the Arduino Nano.

Figure 2.6 shows the circuit diagram of the system.

2.2.3 Program Code

```
/////// Library for LCD16*2
#include <LiquidCrystal.h>
LiquidCrystal lcd(12, 11, 10, 9, 8, 7);
int DC_Volatge_sensor_level=0;
float DC_supply_Voltage_from_input=0;
void setup()

{
  lcd.begin (16, 2);// Initialize LCD 16*2
  lcd.setCursor (0,0);// setcursor on LCD
  lcd.print ("DC Voltage");// print string on LCD
  lcd.setCursor (0,1);// setcursor on LCD
  lcd.print ("Measurement");// print string on LCD
  delay (2000);// set dealy og 2000mSec
```

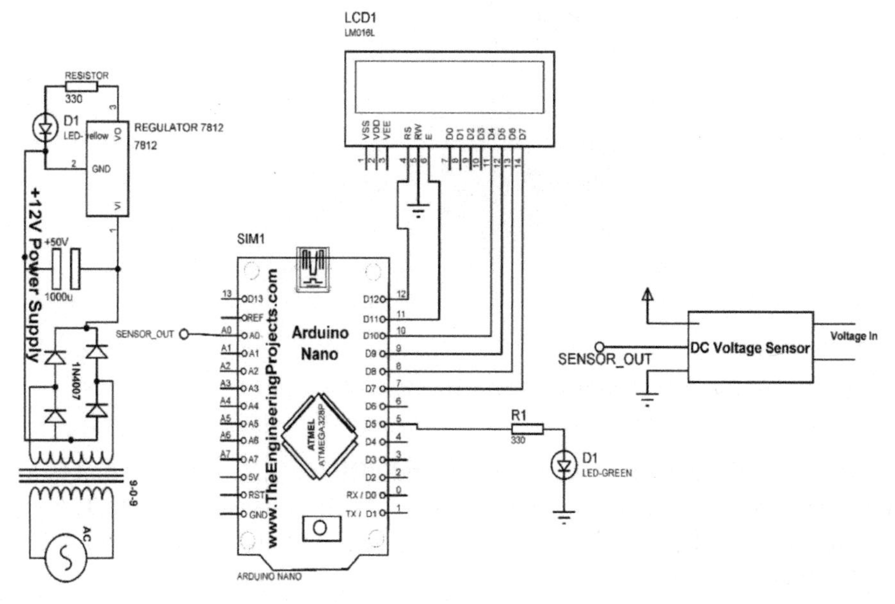

Figure 2.6 Circuit diagram of the system.

```
}

void loop()

{
 DC_Volatge_sensor_level = analogRead(A0);// read analog
    sensor and store in variable
 DC_supply_Voltage_from_input = DC_Volatge_sensor_level *
    (5.0 / 1024.0) * 10;// convert in volatage form
 lcd.clear ();// clear the LCD
 lcd.setCursor (0,0);// set cursor on LCD
 lcd.print ("DC voltage(Vdc):");// print string on LCD
 lcd.setCursor (0,1);// set cursor on LCD
 lcd.print (DC_supply_Voltage_from_input);// print label
    integer on LCD
  delay (300);// set delay of 300mSec
}
}
```

2.3 Serial Communication with RF Modem

This section shows the wireless communication using the RF modem and it is an example of serial communication.

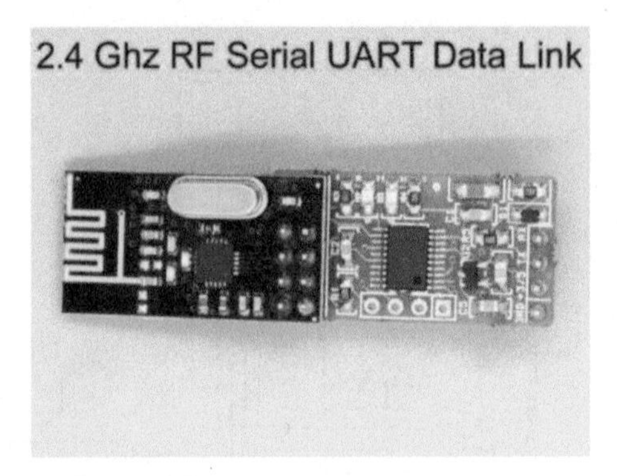

Figure 2.7 2.4-GHz RF serial modem.

The 2.4-GHz modem is an STM8- and nRF24L01-based device able to communicate upto a 50 m range (Figure 2.7). It is capable of transferring serial data over 2.4 GHz RF and supporting bi-directional communication for data logging and sensor reading.

The communication protocol is self-controlled and completely transparent to user interface. When setting an RF serial data communication between microcontrollers or a microcontroller to a PC, the RF modem is most useful and easy to implement. It operates at 5 or 3.3 V and onboard jumper to select the baud rate 9,600 or 115,200. The application of the RF modem includes sensor network and data collection, metering, smart house product, remote control, weather station, etc.

2.3.1 Introduction

Figure 2.8 shows the block diagram of the transmitter section with Arduino NANO and devices like LCD, RF modems, and sensors. It comprises a +12 V/500 mA power supply, an Arduino Nano, an LCD, an MQ6 sensor, and a temperature sensor. The main objective is to measure and display the MQ6 sensor and temperature sensor on LCD and communicate the information data packet through the RF modem via serial communication.

Figure 2.9 shows the block diagram of the receiver section with Arduino NANO and devices like LCD and RF modems. It comprises a +12 V/500 mA power supply, an Arduino Nano, and an LCD. The main objective is to collect the data packet wirelessly from the transmitter and display the data information on the LCD.

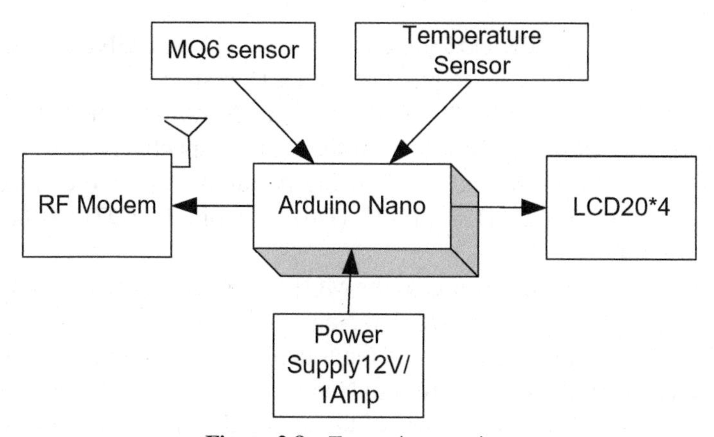

Figure 2.8 Transmitter section.

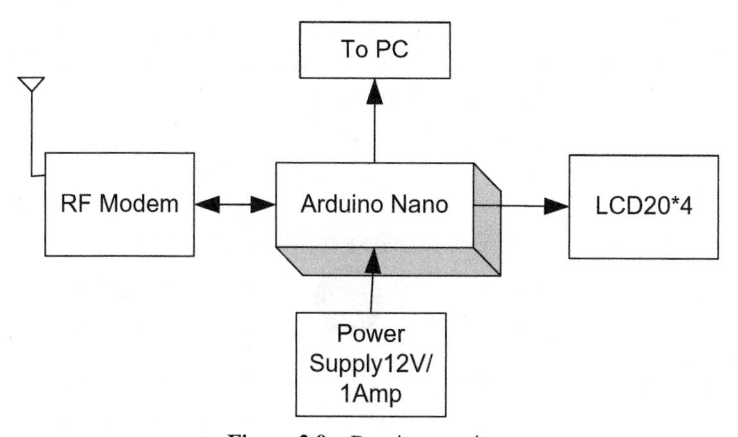

Figure 2.9 Receiver section.

2.3.2 Circuit Diagram

2.3.2.1 Connection of the transmitter

The connections of the transmitter of the given system are as follows:

1. +5V and GND pins of the Arduino Nano are connected to +5V and GND pins of the power supply.
2. Pins 1 and 16 of the LCD are connected to GND of the power supply, respectively.
3. Pins 2 and 15 of the LCD are connected to +5V of the power supply, respectively.

4. Fixed legs of the 10K POT are connected to +5V and GND of the power supply and variable leg to pin 3 of the LCD, respectively.
5. Pin 12, GND, and pin 11 of the Arduino Nano are connected to pin 4(RS), pin 5(RW), and pin 6(E) of the LCD, respectively.
6. Pin 10, pin 9, pin 8, and pin 7 of the Arduino Nano are connected to pin 11 (D4), pin 12(D5), pin 13 (D6), and pin 14 (D7) of the LCD, respectively.
7. +Vcc, GND, and OUT pins of the MQ135 sensor are connected to +5V, GND, and A1 pin of the Arduino Nano, respectively.
8. +Vcc, GND, and OUT pins of the touch sensor are connected to +5V, GND, and A0 pin of the Arduino Nano.
9. +Vcc, GND, TX, and Rx pins of the RF modem are connected to +5V, GND, RX, and TX pin of the Arduino Nano.

Figure 2.10 shows the circuit diagram of the transmitter of the system. Table 2.3 shows the components list for the transmitter section and Table 2.4 shows the components list for receiver section.

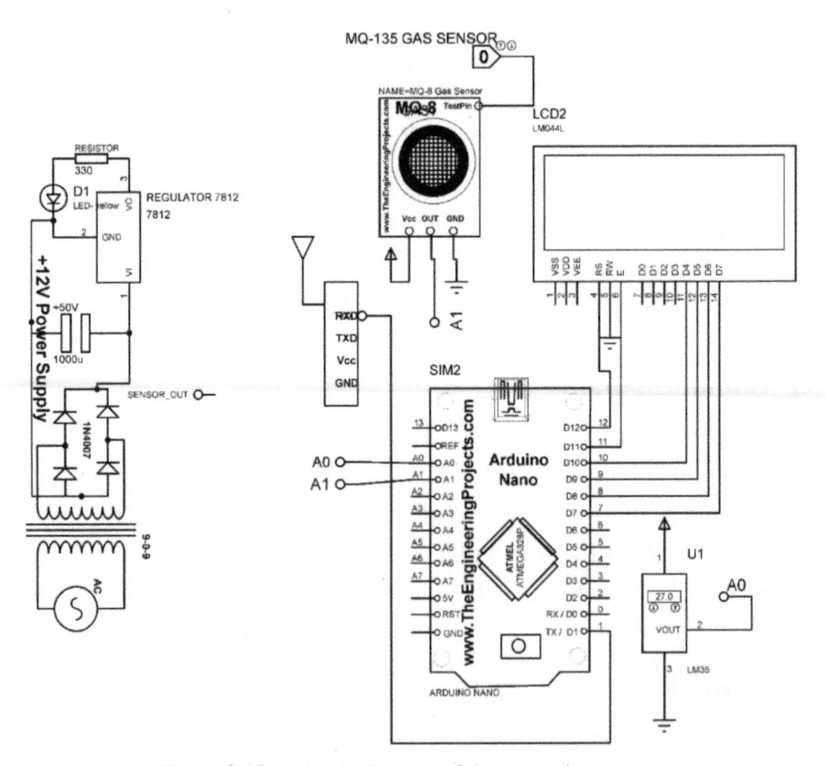

Figure 2.10 Circuit diagram of the transmitter sector.

Table 2.3 Components' list for the transmitter section

Component	Quantity
Power supply 12 V/1 A	1
Arduino Nano	1
Jumper wire M-M	20
Jumper wire M-F	20
Jumper wire F-F	20
Power supply extension (to get more +5V and GND)	1
Level converter to 12 to 5 V, 3.3 V	1
+12V to +5V convertor	1
LCD20*4	1
LCD breakout board/patch	1
RF modem	1
RF modem patch	1
MQ6 sensor	1
LM35 sensor	1

Table 2.4 Components' list for the receiver section

Component	Quantity
Power supply 12 V/1 A	1
Arduino Nano	1
Jumper wire M-M	20
Jumper wire M-F	20
Jumper wire F-F	20
Power supply extension (to get more +5V and GND)	1
Level converter to 12 to 5 V, 3.3 V	1
+12V to +5V convertor	1
LCD20*4	1
LCD breakout board/patch	1
RF modem	1
RF modem patch	1

2.3.2.2 Connections of the receiver

The connections of the receiver of the given system are as follows:

1. +5V and GND pins of the Arduino Nano are connected to +5V and GND pins of the power supply, respectively.
2. Pins 1 and 16 of the LCD are connected to GND of the power supply.
3. Pins 2 and 15 of the LCD are connected to +5V of the power supply.
4. Fixed legs of the 10K POT are connected to +5V and GND of the power supply and variable leg to pin 3 of the LCD, respectively.
5. Pin 12, GND, and pin 11 of the Arduino Nano are connected to pin 4(RS), pin 5(RW), and pin 6(E) of the LCD respectively.

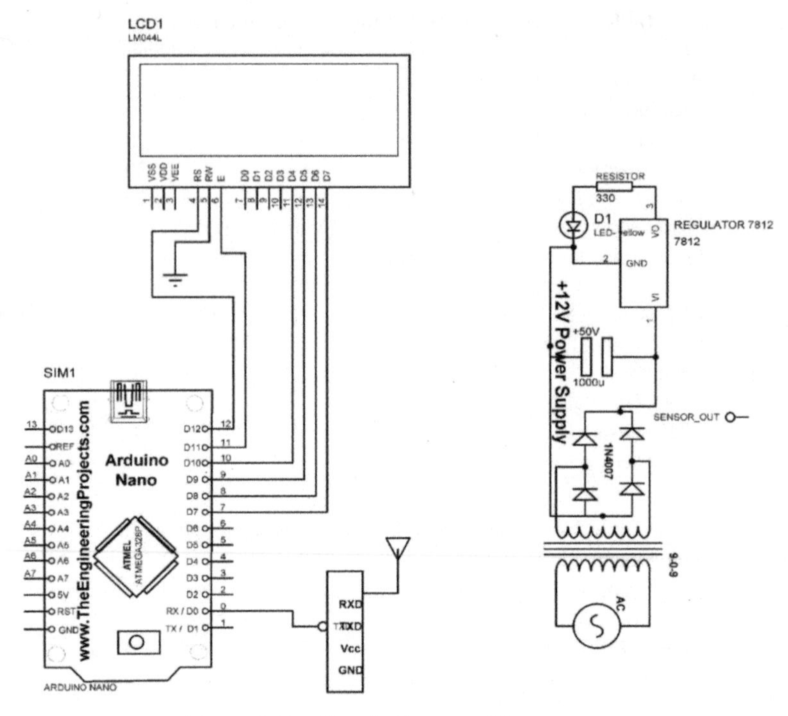

Figure 2.11 Circuit diagram of the receiver.

6. Pin 10, pin 9, pin 8, and pin 7 of the Arduino Nano are connected to pin 11(D4), pin 12(D5), pin 13(D6), and pin 14(D7) of the LCD, respectively.
7. +Vcc, GND, TX, and RX pins of the RF modem are connected to +5V, GND, RX, and TX pins of the Arduino Nano, respectively.

Figure 2.11 shows the circuit diagram of the receiver of the system.

2.3.3 Program Code

This section describes the program of the transmitter and receiver sections.

2.3.3.1 Transmitter Code

```
///////// for Air quality sensor
int MQ135_sensor_pin=A1;
int MQ135_sensor_level=0;
int Touch_sensor_pin=A0;
```

```
int Touch_sensor_state;
#include <LiquidCrystal.h>
LiquidCrystal lcd(12, 11, 10, 9, 8, 7);

void setup()
{
 lcd.begin(20,4);// Initialize LCD
 Serial.begin(9600);// initialize Serial communication
 pinMode(Touch_sensor_pin, INPUT_PULLUP);// make pin A0 as
    active LOW input
}


void loop()
{
 Touch_sensor_state = digitalRead(Touch_sensor_pin);//read
    touch sensor as digital input
 MQ135_sensor_level=analogRead(MQ135_sensor_pin);// read
    MQ135 sensor as analog input
 if(Touch_sensor_state == LOW)// check low state
 {
  lcd.clear();     // clear LCD
  int Touch_sensor_variable=10;
  lcd.setCursor(0,0);// set cursor of LCD
  lcd.print("TOUCH STATUS:");// print string on LCD
  lcd.setCursor(0,1);// set cursor of LCD
  lcd.print("Yes");// print string on LCD
  lcd.setCursor(0,2);// set cursor of LCD
  lcd.print("AIR_Qlty_Level:");// print string on LCD
  lcd.setCursor(0,3);// set cursor of LCD
  lcd.print(MQ135_sensor_level);// print integer on LCD
  Serial.print(Touch_sensor_variable); // send serial data of
     touch sensor state
  Serial.print(":"); // print semicoulumn  as atring
  Serial.print(MQ135_sensor_level);// send serial data of
    MQ135 sensor state
  Serial.print('\r');// print '\r' at last of the sensors
    packet
  delay(50);// set delay of 50mSec
  }

  else
  {
   lcd.clear();    // clear LCD screen
   int Touch_sensor_variable=20;
   lcd.setCursor(0,0);// set cursor on lCD
```

```
    lcd.print("TOUCH STATUS:");// print string on LCD
    lcd.setCursor(0,1);// set cursor on lCD
    lcd.print("Yes");// print string on LCD
    lcd.setCursor(0,2);// set cursor on lCD
    lcd.print("AIR_Qlty_Level:");// print string on LCD
    lcd.setCursor(0,3); set cursor on lCD
    lcd.print(MQ135_sensor_level);// print integer on LCD

    Serial.print(Touch_sensor_variable); // send serial value
     of touch sensor
    Serial.print(":"); // print string as semicolumn
    Serial.print(MQ135_sensor_level);// send serial value of
     MQ135 sensor
    Serial.print('\r');// print '\r' at last of the data
     packet
    delay(50);// set delay of 50mSec
  }
   }
```

2.3.3.2 Receiver Code

```
#include <LiquidCrystal.h>
LiquidCrystal lcd(12, 11, 10, 9, 8, 7);

String inputString_arduino = "";
boolean Arduino_stringComplete = false;

int Touch_sensor_level, MQ135_sensor_level;
void setup()

{
 lcd.begin(20,4);// initialize LCD
 Serial.begin(9600);// initialize Serial communication
 inputString_arduino.reserve(200);// reserve string upto 200
    bytes
}

void loop()
{
 arduino_serialEvent();// call function to read the serial
    data of sensors
 if(Touch_sensor_level == 10) // check the touch sensor state
 {
  lcd.clear();      // clear LCD
  lcd.setCursor(0,0);// set cursor on LCD
  lcd.print("TOUCH STATUS:");// print string on LCD
  lcd.setCursor(0,1);// set cursor on LCD
```

```
  lcd.print("Yes");// print string on LCD
  lcd.setCursor(0,2);// set cursor on LCD
  lcd.print("AIR_Qlty_Level:");// print string on LCD
  lcd.setCursor(0,3);// set cursor on LCD
  lcd.print(MQ135_sensor_level);// print int on LCD
  delay(50);// set delay of 50mSec
}
  else if(Touch_sensor_level == 20)
{
  lcd.clear();
  lcd.setCursor(0,0);// set cursor on LCD
  lcd.print("TOUCH STATUS:");// print string on LCD
  lcd.setCursor(0,1); s//et cursor on LCD
  lcd.print("NO ");// print string on LCD
  lcd.setCursor(0,2);// set cursor on LCD
  lcd.print("AIR_Qlty_Level:");// print string on LCD
  lcd.setCursor(0,3);// set cursor on LCD
  lcd.print(MQ135_sensor_level);// print integer on LCD
  delay(50);// set delay of 50mSec
  }
}

void arduino_serialEvent()
{
 while (Serial.available()>0)// check serial data
 {
  char BYTE_serial = (char)Serial.read();// read serial data
    from RF modem
  inputString_arduino += BYTE_serial;// store data on string
  if (BYTE_serial == '\r')// cheak last byte as termination
    byte of packet
  {
Touch_sensor_level=(((inputString_arduino[0]-48)*10)+((
    inputString_arduino[1]-48)*1));
 MQ135_sensor_level=(((inputString_arduino[3]-48)*100)+((
    inputString_arduino[4]-48)*10)+((inputString_arduino
    [5]-48)*1));/// only 3 byte can be received for MQ135
  }
 }
}
```

3

Interfacing of ESP8266 with Input/Output Devices

This section describes the interfacing among analog sensors, digital sensors, actuators, and serial communication devices.

3.1 Interfacing of ESP8266 with Analog Sensor

3.1.1 Introduction

Figure 3.1 shows the block diagram of the NodeMCU and devices like LCD and analog sensors. It comprises a +12 V/500 mA power supply, a NodeMCU, an LCD, and an LM35 sensor. The main objective is to display the sensory data on the LCD. Table 3.1 shows the components list to develop the system.

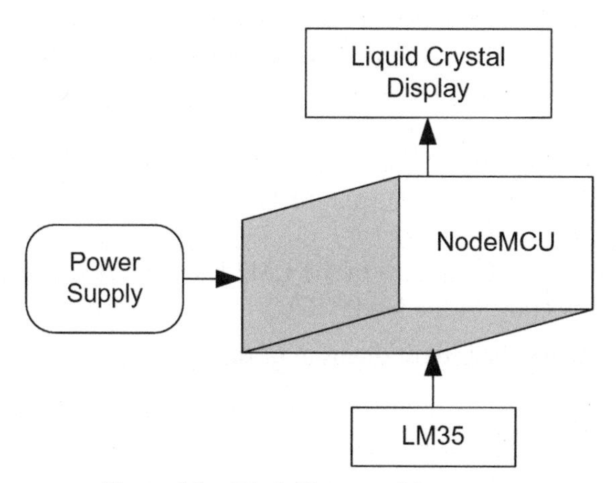

Figure 3.1 Block diagram of the system.

Table 3.1 Components' list

Component	Quantity
Power supply 12 V/1 A	1
NodeMCU	1
Jumper wire M-M	20
Jumper wire M-F	20
Jumper wire F-F	20
Power supply extension (to get more +5V and GND)	1
Level converter to 12 to 5 V, 3.3 V	1
DC voltage sensor	1
+12V to +5V convertor	1
LCD20*4	1
LCD breakout board/patch	1
LM35 sensor	1

3.1.2 Circuit Diagram

The following are the interfacing connections of NodeMCU and the I/O devices:

1. +5V and GND pins of the NodeMCU are connected to +5V and GND pins of the power supply, respectively.
2. Pins 1 and 16 of the LCD are connected to GND of the power supply, respectively.
3. Pins 2 and 15 of the LCD are connected to +5V of the power supply, respectively.
4. Fixed legs of the 10K POT are connected to +5V and GND of the power supply and variable leg to pin 3 of the LCD, respectively.
5. Pin D1, GND, and pin D2 of the NodeMCU are connected to pin 4(RS), pin 5(RW), and pin 6(E) of the LCD, respectively.
6. Pin D3, pin D4, pin D5, and pin D6 of the NodeMCU are connected to pin 11(D4), pin 12(D5), pin 13(D6), and pin 14(D7) of the LCD, respectively.
7. +Vcc, GND, and OUT pins of the LM35 sensor are connected to +5V, GND, and A0 pins of the NodeMCU.

Figure 3.2 shows the circuit diagram of the system.

3.1.3 Program Code

```
#include <LiquidCrystal.h>
const int rs = D1, en = D2, d4 = D3, d5 = D4, d6 = D5,
    d7 = D6;
```

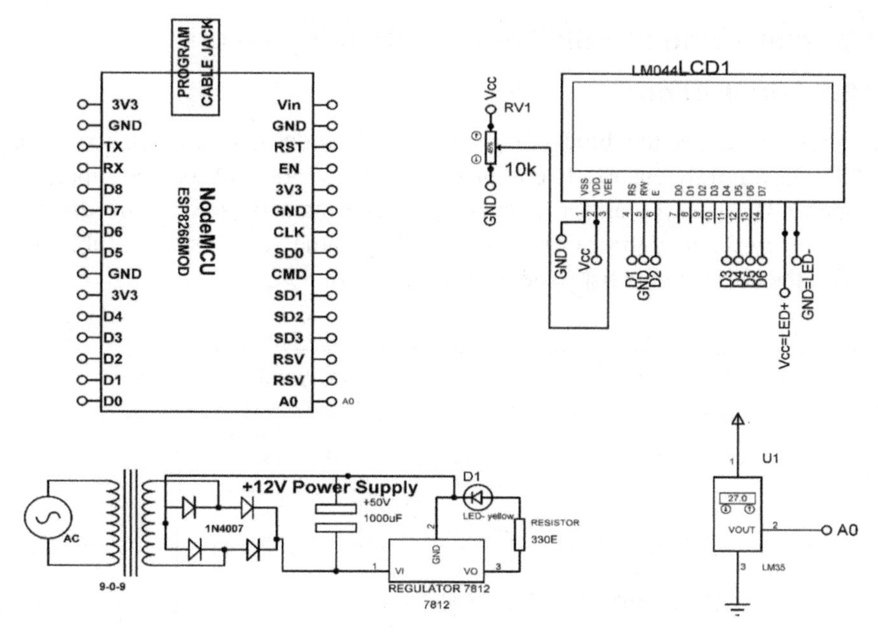

Figure 3.2 Circuit diagram of the system.

```
LiquidCrystal DISPLAY(rs, en, d4, d5, d6, d7);
int LM35_sensor_pin=A0;
void setup()
{
  DISPLAY.begin(16, 2);// initialize the LCD
  DISPLAY.print("Analog_sensor+LCD");// print string on LCD
}


void loop()
{
 int LM35_level=analogRead(LM35_sensor_pin);// read analog
    sensor
 int LM35_ACTUAL=LM35_level/2;// convert it into equivalent
    temperature
 DISPLAY.setCursor(0, 2);// set cursor on LCD
 DISPLAY.print("LM35_level:");// print string on LCD
 DISPLAY.setCursor(0, 2);// set cursor on LCD
 DISPLAY.print(LM35_ACTUAL);// print integer on LCD
 delay(50);// set delay of 50mSec
}
```

3.2 Interfacing of ESP8266 with Digital Sensors

3.2.1 Introduction

Figure 3.3 shows the block diagram of the NodeMCU and devices like LCD and digital sensors. It comprises a +12 V/500 mA power supply, a NodeMCU, an LCD, and a flame sensor. The main objective is to display the sensory data on the LCD by reading the flame sensor status. Table 3.2 shows the components list to develop the system.

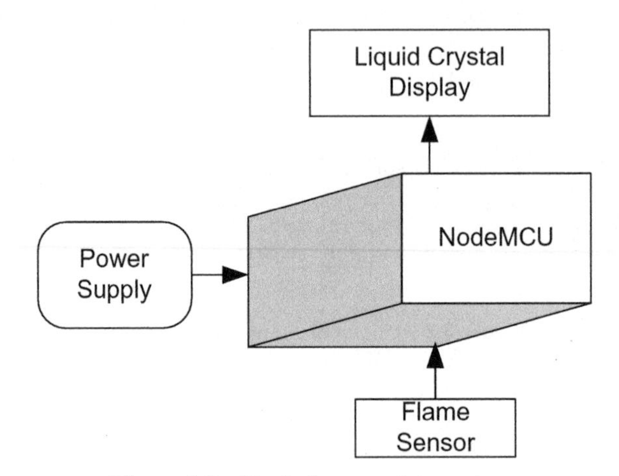

Figure 3.3 Block diagram of the system.

Table 3.2 Components' list

Component	Quantity
Power supply 12 V/1 A	1
NodeMCU	1
Jumper wire M-M	20
Jumper wire M-F	20
Jumper wire F-F	20
Power supply extension (to get more +5V and GND)	1
Level converter to 12 to 5 V, 3.3 V	1
DC voltage sensor	1
+12V to +5V convertor	1
LCD20*4	1
LCD breakout board/patch	1
Flame sensor	1

3.2.2 Circuit Diagram

The following are the interfacing connections of the NodeMCU and the external devices:

1. +5V and GND pins of the NodeMCU are connected to +5V and GND pins of the power supply.
2. Pins 1 and 16 of the LCD are connected to GND of the power supply, respectively.
3. Pins 2 and 15 of the LCD are connected to +5V of the power supply, respectively.
4. Fixed legs of the 10K POT are connected to +5V and GND of power supply and variable leg to pin 3 of the LCD, respectively.
5. Pin D1, GND, and pin D2 of the NodeMCU are connected to pin 4(RS), pin 5(RW), and pin 6(E) of the LCD, respectively.
6. Pin D3, pin D4, pin D5, and pin D6 of the NodeMCU are connected to pin 11(D4), pin 12(D5), pin 13(D6), and pin 14(D7) of the LCD, respectively.
7. +Vcc, GND, and OUT pins of the flame sensor are connected to +5V, GND, and D7 pins of the NodeMCU.

Figure 3.4 shows the circuit schematics of the system to read the digital sensor.

3.2.3 Program Code

```
#include <LiquidCrystal.h>
const int rs = D1, en = D2, d4 = D3, d5 = D4, d6 = D5,
    d7 = D6;
LiquidCrystal DISPLAY(rs, en, d4, d5, d6, d7);
int FLAME_sensor_pin=7;
void setup()
{
 DISPLAY.begin(16, 2);// initialize the LCD
 DISPLAY.print("DIGITAL_sensor+LCD");// print string on LCD
 pinMode(FLAME_sensor_pin,INPUT_PULLUP);// assign pin7 as
    input pin
}

void loop()
{
 LDR_STATUS=digitalRead(FLAME_sensor_pin);// read digital
    sensor
```

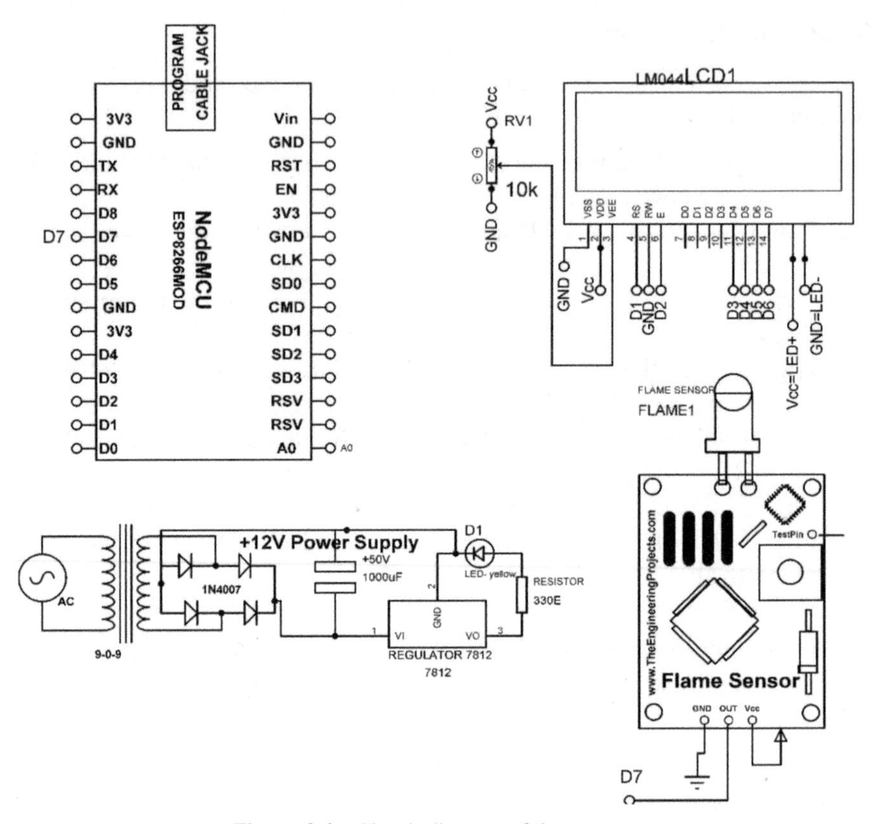

Figure 3.4 Circuit diagram of the system.

```
 if(FLAME_STATUS==LOW)// check status of sensor
 {
  DISPLAY.setCursor(0, 1);// set cursor on LCD
  DISPLAY.print("FLAME STATUS:YES");// print string on LCD
  delay(50);// set delay of 50mSec
 }
 else
 {
  DISPLAY.setCursor(0, 1);// set cursor on LCD
  DISPLAY.print("FLAME STATUS:NO ");// print string on LCD
  delay(50); //set delay of 50mSec
 }
}
```

3.3 NodeMCU and Serial Communication

This section describes the serial communication using NodeMCU.

3.3.1 Introduction

Figure 3.5 shows the block diagram of the NodeMCU and devices like LCD and sensors. It comprises a +12 V/500 mA power supply, a NodeMCU, an LCD, and a flame sensor as digital sensor and an LDR as analog sensor. The main objective is to display the sensory data on the LCD as well as serial communication.

Table 3.3 shows the components' list required to develop the system.

3.3.2 Circuit Diagram

The following are the interfacing connections of the NodeMCU and the external devices:

1. +5V and GND pins of the NodeMCU are connected to +5V and GND pins of the power supply, respectively.
2. Pins 1 and 16 of the LCD are connected to GND of the power supply, respectively.
3. Pins 2 and 15 of the LCD are connected to +5V of the power supply, respectively.
4. Fixed legs of the 10K POT are connected to +5V and GND of the power supply and variable leg to pin 3 of the LCD, repectively.

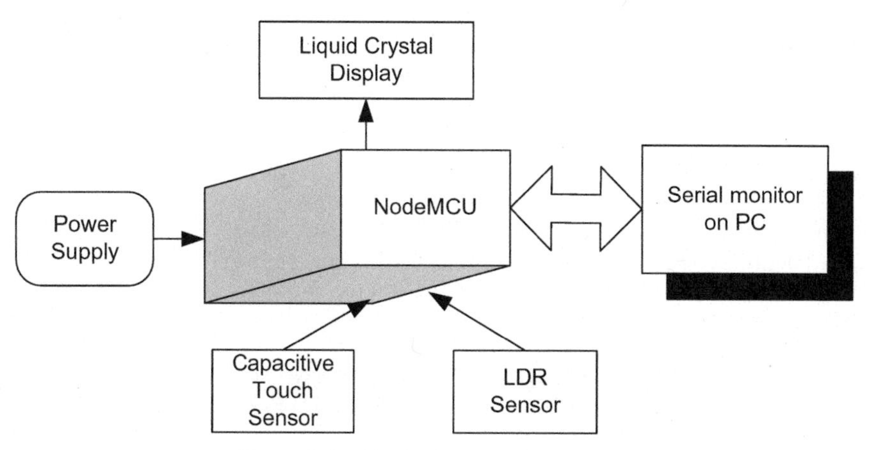

Figure 3.5 Block diagram of the system.

Table 3.3 Components' list

Component	Quantity
Power supply 12 V/1 A	1
NodeMCU	1
Jumper wire M-M	20
Jumper wire M-F	20
Jumper wire F-F	20
Power supply extension (to get more +5V and GND)	1
Level converter to 12 to 5 V, 3.3 V	1
DC voltage sensor	1
+12V to +5V convertor	1
LCD20*4	1
LCD breakout board/patch	1
Capacitive touch sensor	1
LDR sensor	1

5. Pin D1, GND, and pin D2 of the NodeMCU are connected to pin 4(RS), pin 5(RW), and pin 6(E) of the LCD, respectively.
6. Pin D3, pin D4, pin D5, and pin D6 of the NodeMCU are connected to pin 11(D4), pin 12(D5), pin 13(D6), and pin 14(D7) of the LCD, respectively.
7. +Vcc, GND, and OUT pins of the capacitive touch sensor are connected to +5V, GND, and D7 pins of the NodeMCU.
8. +Vcc, GND, and OUT pins of the LDR sensor are connected to +5V, GND, and A0 pins of the NodeMCU.

Figure 3.6 shows the circuit diagram of the system.

3.3.3 Program Code

```
#include <LiquidCrystal.h>

const int rs = D1, en = D2, d4 = D3, d5 = D4, d6 = D5,
      d7 = D6;
LiquidCrystal DISPLAY(rs, en, d4, d5, d6, d7);
int TOUCH_sensor_pin=D7;// connect touch sensor pin on D7 pin
int LDR_sensor_pin=A0;//  onnect LDR on A0 pin
void setup()
       {
        DISPLAY.begin(16, 2);// initialsie LCD
        DISPLAY.print("Analog_sensor+LCD");// print string on
        LCD
```

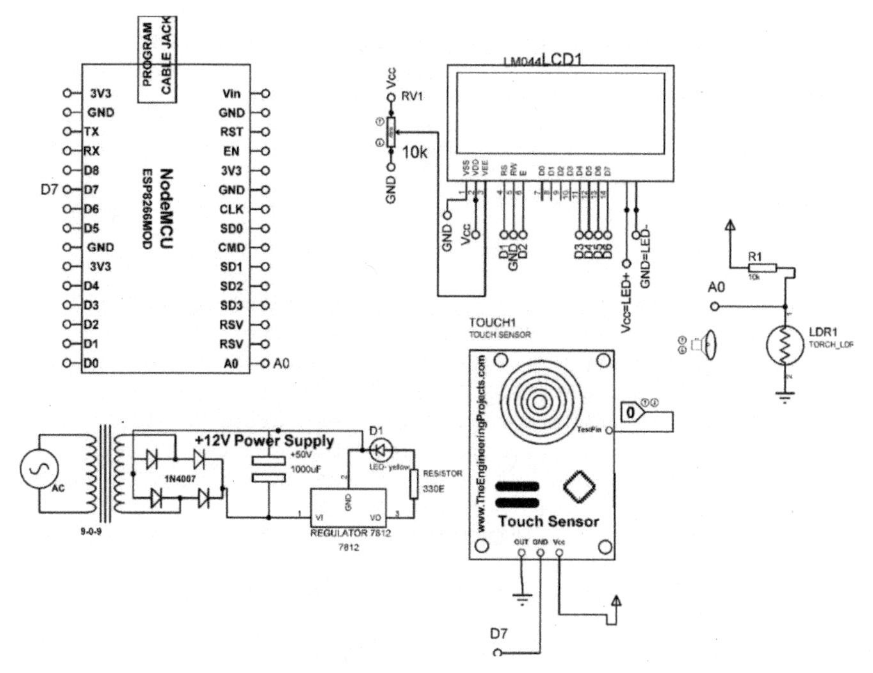

Figure 3.6 Circuit diagram of the system.

```
 pinMode(TOUCH_sensor_pin,INPUT_PULLUP);// assign pin
 D7 as input
 Serial.print(9600);?/ initialize serial communication
}


void loop()
{
 int TOUCH_STATUS=digitalRead(TOUCH_sensor_pin);//
 read touch sensor
 int LDR_LEVEL=analogRead(LM35_sensor_pin);// read LDR
if(TOUCH_STATUS==LOW)// check status
  {
   DISPLAY.setCursor(0, 1);// set cursor on LCD
   DISPLAY.print("TOUCH:YES");// print string on LCD
   DISPLAY.setCursor(0, 2);// set cursor on LCD
   DISPLAY.print("LDR_LEVEL:");// print string on LCD
   DISPLAY.setCursor(5, 2);// set cursor on LCD
   DISPLAY.print(LDR_LEVEL);// print integer on LCD
   DISPLAY.setCursor(8, 2);// set cursor on LCD
```

```
DISPLAY.print(''0C'');// print string on LCD
Serial.println("TOUCH:YES");// print serial data of
Touch sensor
Serial.println(LDR_LEVEL);// print serial data of
LDR
delay(50);
}
else
{
DISPLAY.setCursor(0, 1);// set cursor on LCD
DISPLAY.print("TOUCH:NO ");// print string on LCD
DISPLAY.setCursor(0, 2);// set cursor on LCD
DISPLAY.print("LDR_LEVEL:");// print string on LCD
DISPLAY.setCursor(5, 2);// set cursor on LCD
DISPLAY.print(LDR_LEVEL);// print integer on LCD
DISPLAY.setCursor(8, 2);// set cursor on LCD
DISPLAY.print(''0C'');// print string on LCD
Serial.println("TOUCH:YES");// print serial data of
Touch sensor
Serial.println(LDR_LEVEL);// prit integer value of
LDR sensor
delay(50);// set delay of 50 mSec
}}
```

4

Biometric Car Door Opening System

A car door is used to enter and exit the vehicle which is attached through suitable mechanisms. The door can be opened manually or operated through power electronically.

Hatch or station wagon vehicle bodies generally have either three or five doors. In this case, rear hatch is also considered as a door because it also allows the passengers to entry.

4.1 Introduction

Figure 4.1 shows the block diagram of the NodeMCU and devices like LCD and sensors. It comprises a +12 V/500 mA power supply, a 12 to 5 V convertor, a NodeMCU, an LCD, and a fingerprint sensor. The main objective is to display the fingerprint authentication data on the LCD by matching the fingerprint with pre-stored data inside the memory of the fingerprint sensor. The authentication information with time is uploaded on the cloud server using the NodeMCU/WiFi modem and the information is displayed on mobile App (BLYNK App).

4.2 Circuit Diagram

The following are the interfacing connections of the NodeMCU and the external devices.

1. +5V and GND pins of the NodeMCU are connected to +5V and GND pins of the power supply.
2. Pins 1 and 16 of the LCD are connected to GND of the power supply.
3. Pins 2 and 15 of the LCD are connected to +5V of the power supply.
4. Fixed legs of the 10K POT are connected to +5V and GND of the power supply and variable leg to pin 3 of the LCD.

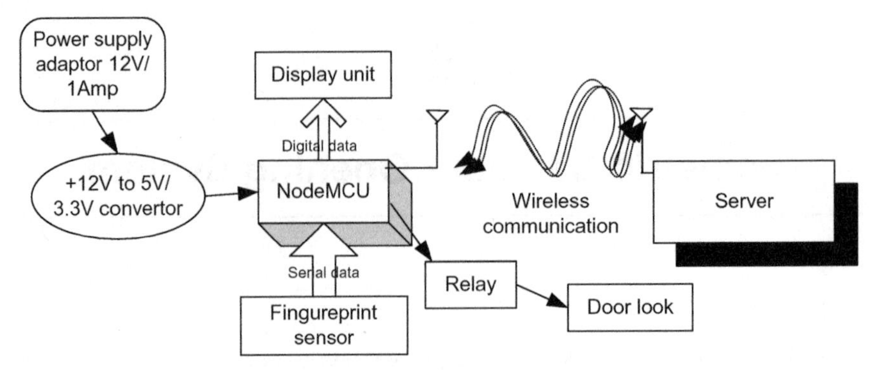

Figure 4.1 Block diagram of the system.

5. Pin D1, GND, and pin D2 of the NodeMCU are connected to pin 4(RS), pin 5(RW), and pin 6(E) of the LCD.
6. Pin D3, pin D4, pin D5, and pin D6 of the NodeMCU are connected to pin 11(D4), pin 12(D5), pin 13(D6), and pin 14(D7) of the LCD.
7. +Vcc, GND, RX_OUT, and SEARCH pins of the fingerprint sensor are connected to +5V, GND, RX and D7 pins of the NodeMCU.

Figure 4.2 shows the circuit diagram of the system.

4.3 Program Code

```
#define BLYNK_PRINT Serial
#include <ESP8266WiFi.h>
#include <BlynkSimpleEsp8266.h>
///// add LCD library
#include <LiquidCrystal.h>
LiquidCrystal lcd(D0, D1, D2, D3, D4, D5);
///// Add softserial library
#include <SoftwareSerial.h>
SoftwareSerial mySerial_one(6,7); // 6 rx /7 tx for
    Fingerprint
///// add credentials
char auth[] = "8507cac915f04a1bb4b00987e420afa0";//
    authentication token
char ssid[] = "ESPServer_RAJ";// id of hotspot
char pass[] = "RAJ@12345";// password of hotspot
BlynkTimer timer;
int search_pin=5;// attach search pin of fingerprint sensor
    to pin 5
```

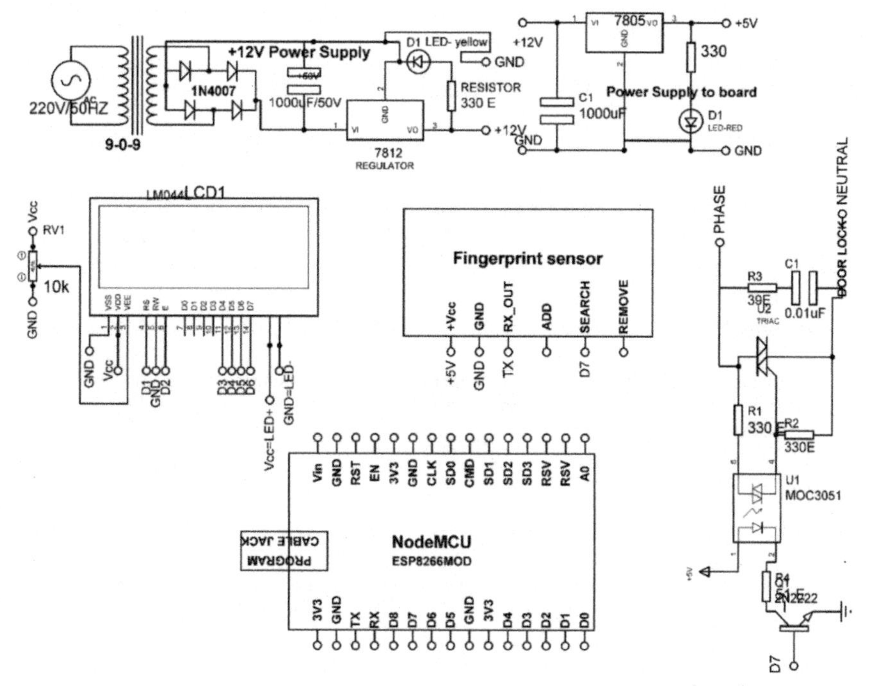

Figure 4.2 Circuit diagram of the system.

```
int digitallockpin=4;// attach digital lock pin to pin 4
WidgetLCD LCD_BLYNK(V0); // configure blynk LCD on virtual
    pinV0

void READ_SENSOR()// function to read the sensor

{
 digitalWrite(search_pin,LOW);/// make serach pin low
 delay(50);// set delay of 50mSec
 digitalWrite(search_pin,HIGH);/// make serach pin high
 int X= mySerial_one.read();// record serial data on X
 if(X==0)// check X
 {
 lcd.setCursor(0,1);// set cursor of LCD
 lcd.print("FIRST          ");// print string on LCD
 LCD_BLYNK.print(0,1,"FIRST              ");// print string
    on blynk LCD
 digitalWrite(digitallockpin,HIGH);// make digital lock pin
    high
```

```
Blynk.virtualWrite(V5,X);// send data to virtual pin V5 of
   blynk
delay(20);// set delay of 50mSec
}
if(X==1)
{
lcd.setCursor(0,1);// setcursor on LCD
lcd.print("SECOND              ");// print string on LCD
LCD_BLYNK.print(0,1,"SECOND                      "); print string
   on blynk LCD
digitalWrite(digitallockpin,HIGH);// make digital lock pin
   to HIGH
Blynk.virtualWrite(V5,X);// send data to virtual pin V5 of
   blynk
delay(20);// set delay of 50mSec
}
 if(X==2)
{
lcd.setCursor(0,1);// setcursor on LCD
lcd.print("THIRD              ");// print string on LCD
LCD_BLYNK.print(0,1,"THIRD                      "); //print string
   on blynk LCD
digitalWrite(digitallockpin,HIGH); //make digital lock pin
   to HIGH
Blynk.virtualWrite(V5,X);// send data to virtual pin V5 of
   blynk
delay(20);// set delay of 20mSec
}
 if(X==3)
{
lcd.setCursor(0,1);// set cursor on LCD
lcd.print("FOUR               ");// print string on LCD
LCD_BLYNK.print(0,1,"FOUR                        ");// print string
   on blynk LCD
digitalWrite(digitallockpin,HIGH);// make digital lock pin
   to HIGH
Blynk.virtualWrite(V5,X);// send data to virtual pin V5 of
   blynk
delay(20);// set delay of 20mSec
}
if(X==4)
{
lcd.setCursor(0,1);// set cursor on LCD
lcd.print("FIVE                 ");// print string on LCD
LCD_BLYNK.print(0,1,"FIVE                      ");// print string on
    blynk LCD
```

```
 digitalWrite(digitallockpin,HIGH);// make digital lock pin
    to HIGH
 Blynk.virtualWrite(V5,X);// send data to virtual pin V5 of
    blynk
 delay(20);// set delay of 50mSec
 }
  if(X==5)
 {
 lcd.setCursor(0,1); //set cursor on LCD
 lcd.print("SIX           ");// print string on LCD
 LCD_BLYNK.print(0,1,"SIX            ");// print string on
    blynk LCD
 digitalWrite(digitallockpin,HIGH);// make digital lock pin
    to HIGH
 Blynk.virtualWrite(V5,X);// send data to virtual pin V5 of
    blynk
 delay(20);// set delay of 20mSec
 }
  if(X==0xFF)
 {
 lcd.setCursor(0,1);// send data to virtual pin V5 of blynk
 lcd.print("Punch The Fingure.. ");// print string on LCD
 LCD_BLYNK.print(0,1,"PUNCH THE FINGURE      ");// print
    string on blynk LCD
 digitalWrite(digitallockpin,LOW); //make digital lock pin to
     LOW
 Blynk.virtualWrite(V5,X);// send data to virtual pin V5 of
    blynk
 delay(20);// set delay of 20mSec
 }

}

void setup()
{
 Serial.begin (9600); // initialize serial communication for
    GPS
 mySerial_one.begin (9600);//  initialize for finger print
    serial baud rate
 lcd.begin(20, 4); // initialize LCD
 pinMode(search_pin,OUTPUT);/// set search pin as input pin
 pinMode(digitallockpin,OUTPUT);///// set digital lock pin as
     output pin
 lcd.setCursor(0,0);// set cursor of LCD
 lcd.print("welcome");// print string on LCD
 Blynk.begin(auth, ssid, pass);// initialize blynk app
```

```
 timer.setInterval(10000L,READ_SENSOR);//// set sample rate
    of read sensor function 10 sec
 delay(2000);// set delay of 2000mSec

}
void loop()
{
Blynk.run();//initialize bynk
timer.run(); // Initialize BlynkTimer
}
```

4.4 Blynk APP

Blynk is iOS and Android platform to design mobile apps. To design the app, download the latest Blynk library from: https://github.com/blynkkk/blynk-library/releases/latest

Mobile App can easily be designed just by dragging and dropping widgets on the provided space. Tutorials can be downloaded from: http://www.blynk.cc

Steps to design Blynk App

1. **Step 1:** Download and install the Blynk App for your mobile Android or iphone from http://www.blynk.cc/getting-started/
2. **Step 2:** Create a Blynk Account
3. **Step 3:** Create a new project
   Click on + for creating a new project and choose the theme dark (black background) or light (white background) and click on create (Figure 4.3).
4. **Step 4:** Auth token is a unique identifier, which will be received on the email address of the user provided at the time of making the account. Save this token, as this is required to copy in the main program of the receiver section.
5. **Step 5:** Select the device to which smart phone needs to communicate, e.g., ESP8266 (NodeMCU).
6. **Step 6:** Open widget box and select the components required for the project. For this project, five buttons are selected.
7. **Step 7:** Tap on the widget to get its settings and select virtual terminals as V1 and V2 for each button, which needs to be defined later on the program.
8. **Step 8:** After completing the widget settings, run the project.
9. Front end of the APP for the proposed system (Figure 4.4).

Figure 4.3 Create a new project.

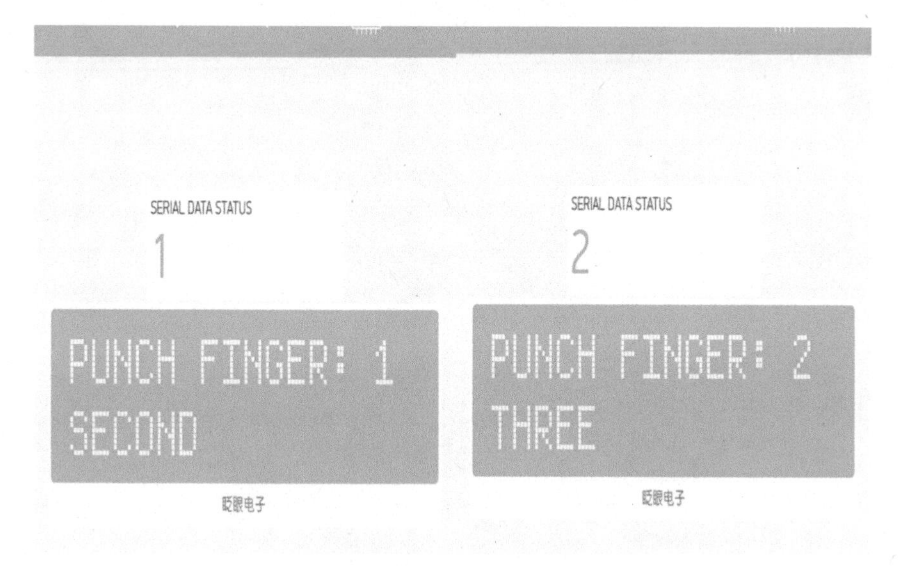

Figure 4.4 BLYNK APP.

5

Accident Monitoring System

Vehicle population is growing very rapidly compared to the economy and population growth of country and population growth. This is due to the rapid growth of technology and its infrastructure, available due to globalization. This is a very positive sign on one side but another side this advent of tech also increases the road traffic hazards, which leads to a lot of irreparable loss. Road safety measures need to improve as well as immediate attention is required at accidents-prone places, which can save the lives of people who are victims of accidents.

5.1 Introduction

Figure 5.1 shows the block diagram of the NodeMCU and devices like LCD and sensors. It comprises a +12 V/500 mA power supply, a 12 to 5 V convertor, an Arduino Nano, a NodeMCU, an LCD, and a pressure sensor. The main objective is to display the data of the pressure sensor and location by GPS coordinates on the LCD by reading the bump or hit on the car. The sensory information and location by GPS coordinates transfer to the Node MCU via a serial link. The information with time can be uploaded to the cloud server using the NodeMCU/WiFi modem. Table 5.1 shows the list of components to develop the system.

5.2 Circuit Diagram

The following are the interfacing connections of the NodeMCU and the external devices.

1. +5V and GND pins of the NodeMCU and Arduino Nano are connected to +5V and GND pins of the power supply.
2. Pins 1 and 16 of the LCD are connected to GND of the power supply.

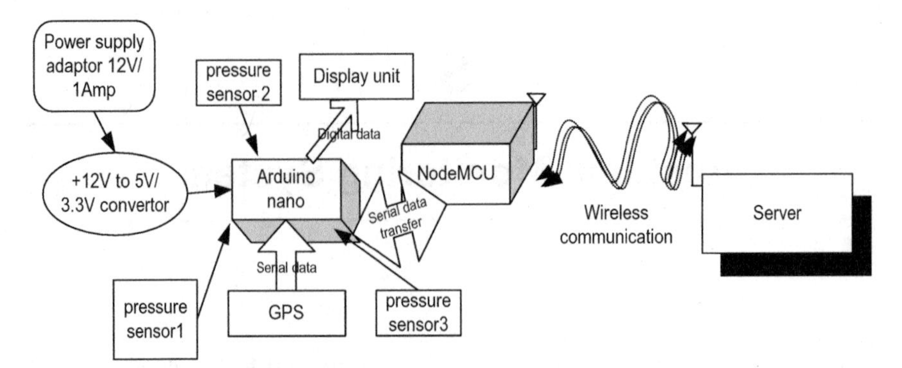

Figure 5.1 Block diagram of the system.

Table 5.1 Components' list

Component	Quantity
Power supply 12 V/1 A	1
NodeMCU	1
Arduino Nano	1
Jumper wire M-M	20
Jumper wire M-F	20
Jumper wire F-F	20
Power supply extension (to get more +5V and GND)	1
DC voltage sensor	1
+12V to +5V convertor	1
LCD20*4	1
LCD breakout board/patch	1
Pressure sensor	3
GPS	1

3. Pins 2 and 15 of the LCD are connected to +5V of the power supply.
4. Fixed legs of the 10K POT are connected to +5V and GND of the power supply and variable leg to pin 3 of the LCD.
5. Pin 12, GND, and pin 11 of the Arduino are connected to pin 4(RS), pin 5(RW), and pin 6(E) of the LCD.
6. Pin 10, pin 9, pin 8, and pin7 of the Arduino are connected to pin 11(D4), pin 12(D5), pin 13(D6), and pin 14(D7) of the LCD, respectively.
7. +Vcc, GND, and RX_OUT pins of the GPS are connected to +5V, GND, and RX pins of the Arduino.
8. +Vcc, GND, and OUT pins of pressure sensor 1 are connected to +5V, GND, and A0 pins of the Arduino.

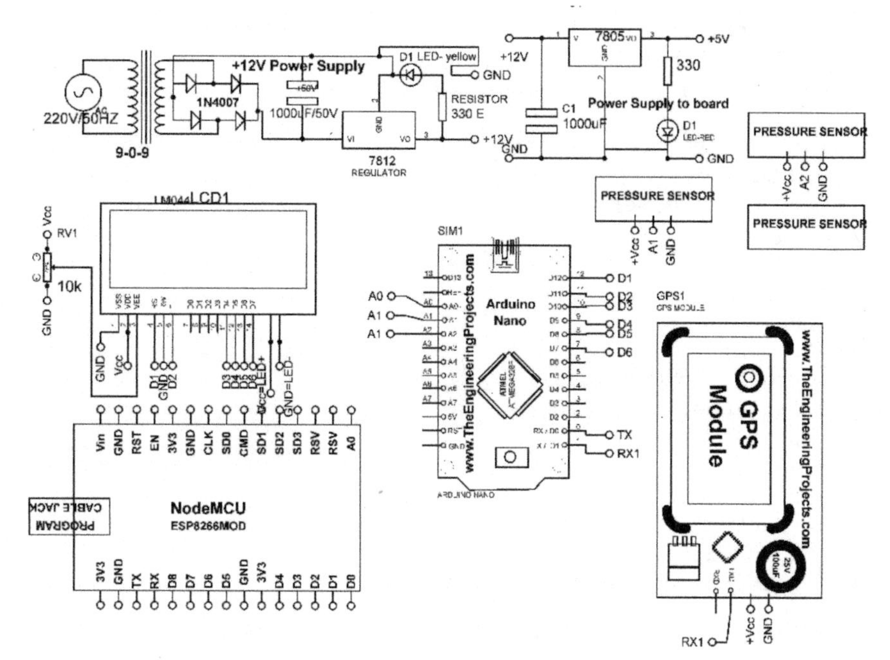

Figure 5.2 Circuit diagram of the system.

9. +Vcc, GND, and OUT pins of pressure sensor 2 are connected to +5V, GND, and A1 pins of the Arduino.
10. +Vcc, GND, and OUT pins of pressure sensor 3 are connected to +5V, GND, and A2 pins of the Arduino.
11. The TX pin of the Arduino is connected to the RX pin of the NodeMCU.

Figure 5.2 shows the circuit diagram of the system.

5.3 Program Code

Sections 5.3.1 and 5.3.2 show the program code for Arduino nano and NodeMCU, respectively.

5.3.1 Program Code for Arduino Nano

```
#include <TinyGPS.h>// add header of GPS
#include <LiquidCrystal.h>// add header for LCD
```

```
LiquidCrystal lcd(13, 12, 11, 10, 9, 8);// pins connection of
     LCD with ardiuino


#include <SoftwareSerial.h>// softserial library to make
    another serial port
SoftwareSerial mySerial_one(6,7); // 6 as RX pin  /7 as TX
    pin


//// GPS
TinyGPS gps;
void getgps(TinyGPS &gps);
float latitude, longitude;
byte a;
 void getgps(TinyGPS &gps)
{
 float latitude, longitude;
 decode and display position data
 gps.f_get_position(&latitude, &longitude);


 lcd.setCursor (0,3);// set cursor of LCD
 lcd.print ("Lat:");// print string on LCD
 lcd.print (latitude,5);// print latitude value on LCD
 Serial.print (latitude);// serial print the value of
    latitude
 Serial.print ("  ");// set gap
 lcd.print (" ");// print on lCD

 lcd.setCursor (10,3);// set cursor of LCD
 lcd.print ("Long:");// print string on LCD
 lcd.print (longitude,5);// print longitude value on LCD
 Serial.println (longitude);// serial print the value of
    longitude
 lcd.print (" "); print string on LCD
 delay(3000); // wait for 3 seconds
}
void  CALL_GPS()
{
 byte a;
 if ( Serial.available() > 0 ) // if there is data coming
    into the serial line
 {
  a = Serial.read(); // get the byte of data
 if(gps.encode(a)) // if there is valid GPS data...
```

```
  {
   getgps(gps); // grab the data and display it on the LCD
     }
  }

   }


void setup()
{
 Serial.begin(9600); // for GPS
 mySerial_one.begin(9600);//  for finger print
 lcd.begin(20, 4); // initialize LCD
 lcd.setCursor(0,0);// set cursor on LCD
 lcd.print("welcome to");// print string on LCD
 lcd.setCursor(0,0);// set cursor on LCD
 lcd.print("GPS Monitoring");// print string on LCD
 delay(2000);// wait for 2 sec
 lcd.clear();// clear the LCD contents

}

void loop()
{
 int pressure_sensor1=analogRead(A0);// read analog pressure
    sensor 1
 int pressure_sensor2=analogRead(A1);// read analog pressure
    sensor 2
 int pressure_sensor3=analogRead(A2);// read analog pressure
    sensor 3
 CALL_GPS(); // call GPS function for longitude and latitude
 lcd.setCursor(0,0);// set cursor on LCD
 lcd.print("P0:");// print string on LCD
 lcd.setCursor(3,0);// set cursor on LCD
 lcd.print(pressure_sensor1);// print int value of pressure
    sensor 1
 lcd.setCursor(10,0);// set cursor on LCD
 lcd.print("P1:");// print string on LCD
 lcd.setCursor(13,0);// set cursor on LCD
 lcd.print(pressure_sensor2);// print int value of pressure
    sensor 2
 lcd.setCursor(0,1);// set cursor on LCD
 lcd.print("P2:");// print string on LCD
 lcd.setCursor(3,1);// set cursor on LCD
 lcd.print(pressure_sensor3);// print int value of pressure
    sensor 3
```

```
Serial.print(pressure_sensor1);// send pressure sensor 1
   value on serial port
Serial.print(";");// print string on LCD
Serial.print(pressure_sensor2);// send pressure sensor 2
   value on serial port
Serial.print(";");//print string on LCD
Serial.print(pressure_sensor3);// send pressure sensor 3
   value on serial port
Serial.print(";");//print string on LCD
Serial.print(latitude);// send int value of latitude on
   serial port
Serial.print(";");//print string on LCD
Serial.print(longitude);// send int value of latitude on
   serial port
Serial.print('\n');// print '\n' on LCD
}}
```

5.3.2 Program Code for NodeMCU

```
#include <SoftwareSerial.h>
#include <ESP8266WiFi.h>
#include "StringSplitter.h"
SoftwareSerial mySerial(D7,D8,false,256);// make D7 and D8 as
    RX and TX pin
String apiKey1 = "O44YTW0Z5WNO17N8";// api key from thinspeak
const char* ssid = "ESPServer_RAJ";// hotspot ID
const char* password = "RAJ@12345";// hotspot password
const char* server = "api.thingspeak.com";
WiFiClient client;
String PRESS_SENSOR1,PRESS_SENSOR2,PRESS_SENSOR3,latitude,
   longitude;
String inputString_NODEMCU = "";         // a string to hold
   incoming data

   void setup()
   {
   Serial.begin(115200);// initialize serial communication
   mySerial.begin(115200);// initialize serial communication
    of D7 and D8 pins
   inputString_NODEMCU.reserve(200);// reserve bytes
   delay(10);// wait for 10mSec
   WiFi.begin(ssid, password);// initialize wiFi
   communication
   Serial.println();// print serial
   Serial.println();// print serial
   Serial.print("Connecting to ");// print string on LCD
```

```
    Serial.println(ssid);// print ssid

    while (WiFi.status() != WL_CONNECTED)// check wifi status
    {
    delay(500);// wait for 500mSec
    Serial.print(".");// print serial
    }
    Serial.println("");// print serial
    Serial.println("WiFi connected");// print string on
    serial
    }

     void loop()
     {


      if (client.connect(server,80))// check server
           {
     serialEvent_NODEMCU();// call function to read serial
    data from nodeMCU
     send1_TX_ACCIDENT_PARA();// Call function to send data
    on thingspeak server
           }
         client.stop();
         Serial.println("Waiting");
         delay(20000);// thingspeak needs minimum 15 sec
    delay between updates
     }
void send1_TX_ACCIDENT_PARA()// function to send data on
    thingspeak server
{
    String postStr = apiKey1;
    postStr +="&field1=";
    postStr += String(PRESS_SENSOR1);// data on field 1
    postStr +="&field2=";
    postStr += String(PRESS_SENSOR2);// data on field 2
    postStr +="&field3=";
    postStr += String(PRESS_SENSOR3);// data on field 3
    postStr +="&field4=";
    postStr += String(latitude);// data on field 4
    postStr +="&field5=";
    postStr += String(longitude);// data on field 5
    postStr += "\r\n\r\n";
    client.print("POST /update HTTP/1.1\n");
    client.print("Host: api.thingspeak.com\n");
    client.print("Connection: close\n");
```

```
    client.print("X-THINGSPEAKAPIKEY: "+apiKey1+"\n");
    client.print("Content-Type: application/x-www-form-
    urlencoded\n");
    client.print("Content-Length: ");
    client.print(postStr.length());
    client.print("\n\n");
    client.print(postStr);
    Serial.print("Send data to channel-1 ");// send string on
     serial
    Serial.print("Content-Length: ");// send string on serial
    Serial.print(postStr.length());// print length of string
    Serial.print("Field-1: ");// send string on serial
    Serial.print(PRESS_SENSOR1);// print sensor value 1 on
    serial
    Serial.print("Field-2: ");// send string on serial
    Serial.print(PRESS_SENSOR2);// print sensor value 1 on
    serial
    Serial.print("Field-3: ");// send string on serial
    Serial.print(PRESS_SENSOR3);// print sensor value 1 on
    serial
    Serial.print("Field-4: ");// send string on serial
    Serial.print(latitude);// print latituder value  on
    serial
    Serial.print("Field-5: ");// send string on serial
    Serial.print(longitude);// print lonitude value on serial
    Serial.println(" data send");// send string on serial
}


void serialEvent_NODEMCU() // function to read serial data
    form Arduino Nano
{
 while (mySerial.available()>0)// check serial data on RX pin
 {
 inputString_NODEMCU = mySerial.readStringUntil('\n');// Get
    serial input from Arduino nano


 StringSplitter *splitter = new StringSplitter(
    inputString_NODEMCU, ',', 5);  // use
 string splitter to separate the data from arduono nano
  StringSplitter(string_to_split, delimiter, limit)
   int itemCount = splitter->getItemCount();


   for(int i = 0; i < itemCount; i++)
```

```
    {
     String item = splitter->getItemAtIndex(i);
     PRESS_SENSOR1 = splitter->getItemAtIndex(0);// sensor1
    data
     PRESS_SENSOR2 = splitter->getItemAtIndex(1);// sensor1
    data
     PRESS_SENSOR3 = splitter->getItemAtIndex(2);// sensor1
    data
     latitude= splitter->getItemAtIndex(3);// latitude data
     longitude=splitter->getItemAtIndex(4);// longitude data
    }
    inputString_NODEMCU = "";// blank the string again
    delay(200);// wait for 200 mSec
    }
   }
```

5.4 ThingSpeak Server

1. Sign In to ThingSpeak by creating a new MathWorks account.
2. Click Channels > MyChannels (Figure 5.3).
3. Click New Channel (Figure 5.4).

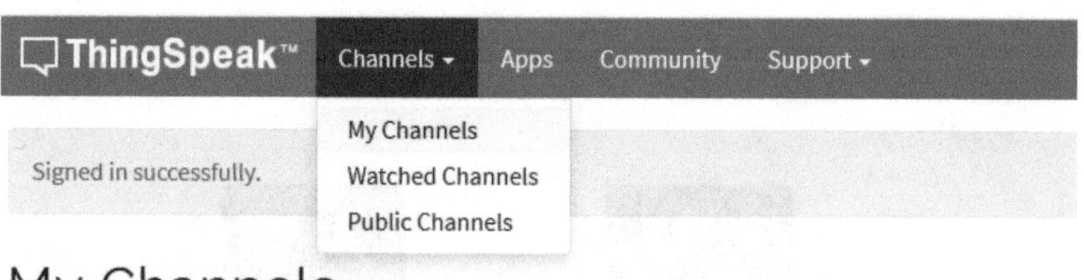

Figure 5.3 Window for ThingSpeak.

Name	Created	Updated At
Channel 293693 Private | Public | Settings | Sharing | API Keys | Data Import / Export	2017-06-26	2017-09-20 04:23

Figure 5.4 New channel in my channels.

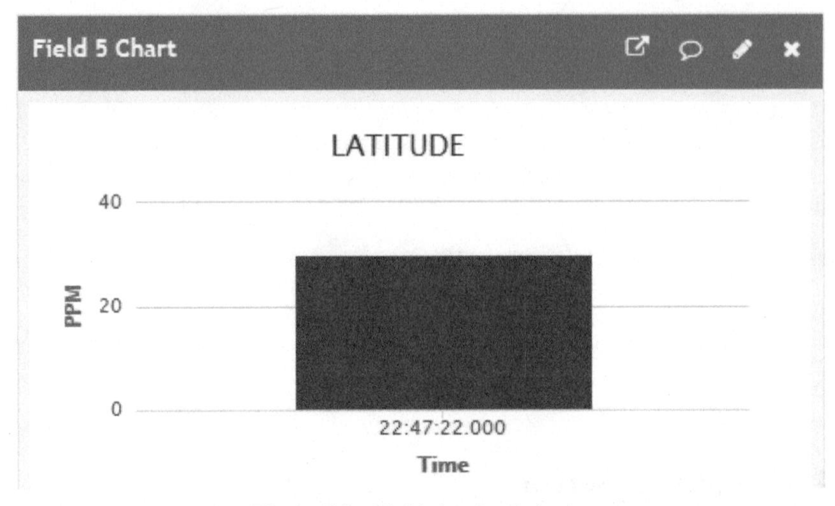

Figure 5.5 Field showing latitude.

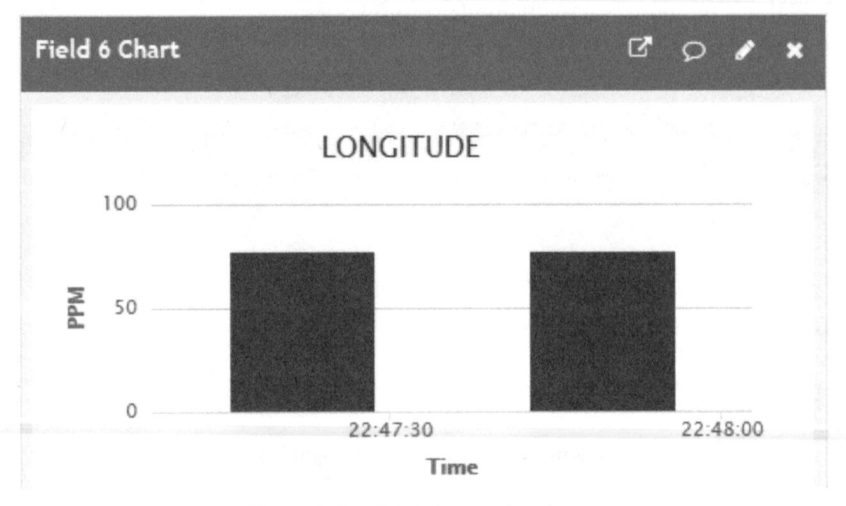

Figure 5.6 Field showing longitude.

4. Check the boxes next to Fields 1–1. Enter the channel setting values. Click Save Channel at the bottom of the settings.
5. Check API write key (this key needs to write in the program for local server).
6. Fields will show the sensory data in the form of graphs (Figures 5.5–5.9).

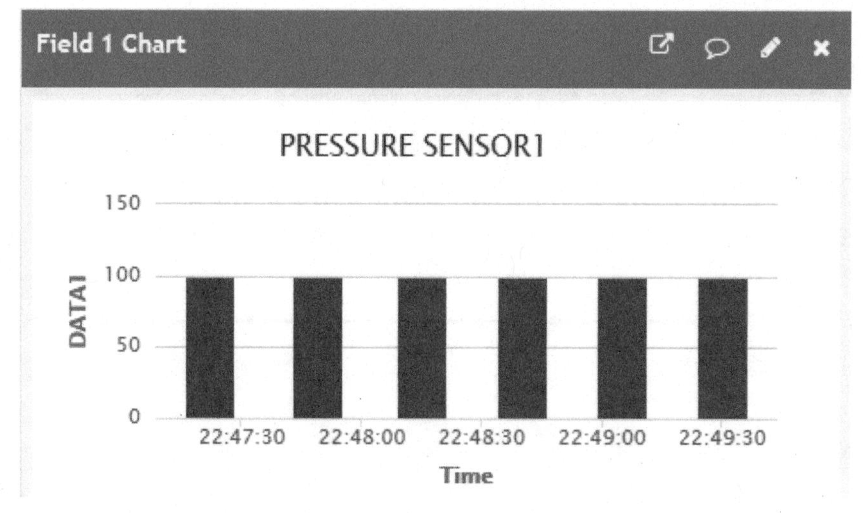

Figure 5.7 Field showing pressure sensor 1(mb).

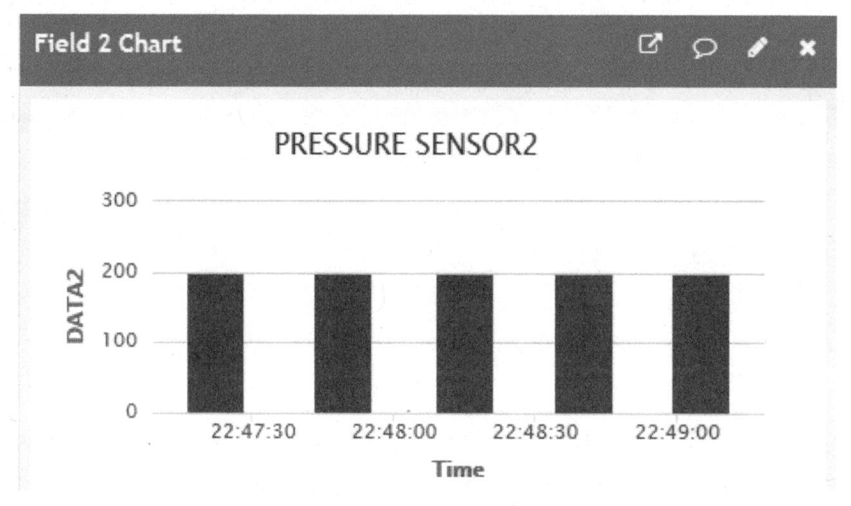

Figure 5.8 Pressure sensor 2 (mb).

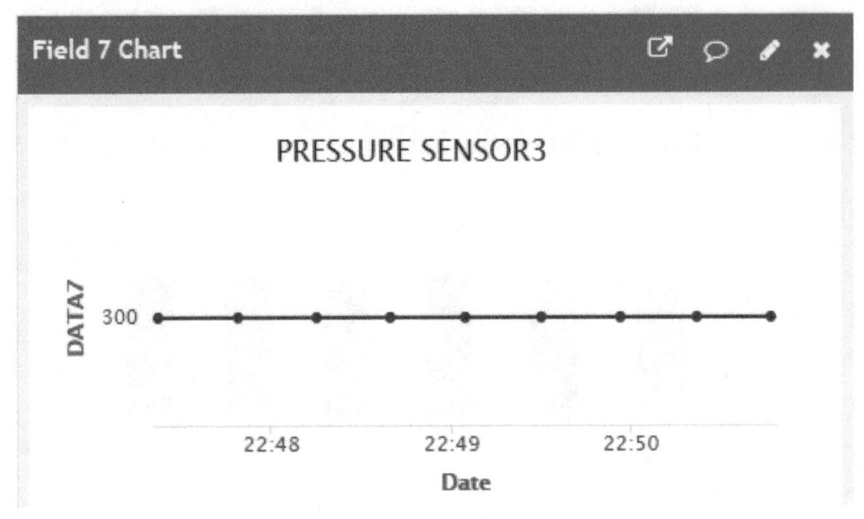

Figure 5.9 Pressure sensor 3 (mb).

6

Engine Oil and Coolant Level Monitoring System

In automobiles, one of the most neglected areas is engine oil and coolant levels. These two fluids play a vital role in any internal combustion engine. It is useful to display levels so that users can aware about its condition before it goes too low. The system is proposed for engine oil and coolant monitoring and display.

6.1 Introduction

Figure 6.1 shows the block diagram of the system with Arduino Nano, NodeMCU, and its other devices. It comprises a +12 V/500 mA power supply, a 12 to 5 V convertor, an Arduino Nano, a NodeMCU, an LCD, and level sensors. The main objective is to display the level of engine oil and coolant on the LCD by reading it through sensors. The sensory information from the Arduino Nano is transferred to the Node MCU via a serial link. The information with time can be uploaded to the cloud server using the NodeMCU/WiFi modem. Table 6.1 shows the list of components to develop the system.

6.2 Circuit Diagram

The following are the interfacing connections of the NodeMCU and the external devices.

1. +5V and GND pins of the NodeMCU and Arduino Nano are connected to +5V and GND pins of the power supply.
2. Pins 1 and 16 of the LCD are connected to GND of the power supply.
3. Pins 2 and 15 of the LCD are connected to +5V of the power supply.

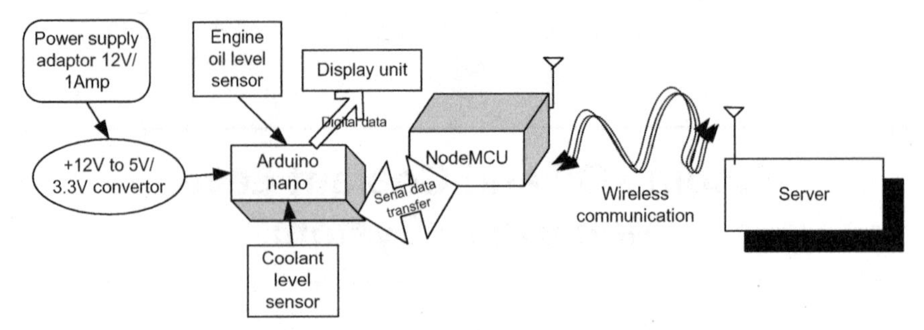

Figure 6.1 Block diagram of the system.

Table 6.1 Components' list

Component	Quantity
Power supply 12 V/1 A	1
NodeMCU	1
Arduino Nano	1
Jumper wire M-M	20
Jumper wire M-F	20
Jumper wire F-F	20
Power supply extension (to get more +5V and GND)	1
DC voltage sensor	1
+12V to +5V/ 3.3 V convertor	1
LCD20*4	1
LCD breakout board/patch	1
level sensors	2

4. Fixed legs of the 10K POT are connected to +5V and GND of the power supply and variable leg to pin 3 of the LCD.
5. Pin D1, GND, and pin D2 of the Arduino Nano are connected to pin 4(RS), pin 5(RW), and pin 6(E) of the LCD, respectively.
6. Pin D3, pin D4, pin D5, and pin D6 of the Arduino Nano are connected to pin 11(D4), pin 12(D5), pin 13(D6), and pin 14(D7) of the LCD, respectively.
7. +Vcc, GND, TRIG, ECHO, and OUT pins of the engine oil level are connected to +5V, GND, and pins 6 and 5 of the Arduino Nano.
8. +Vcc, GND, and SERIAL_OUT pins of the coolant level sensor are connected to +5V, GND, and RX pins of the Arduino Nano.
9. The TX pin of the Arduino Nano is connected to the RX pin of the NodeMCU.

Figure 6.2 shows the circuit diagram of the system.

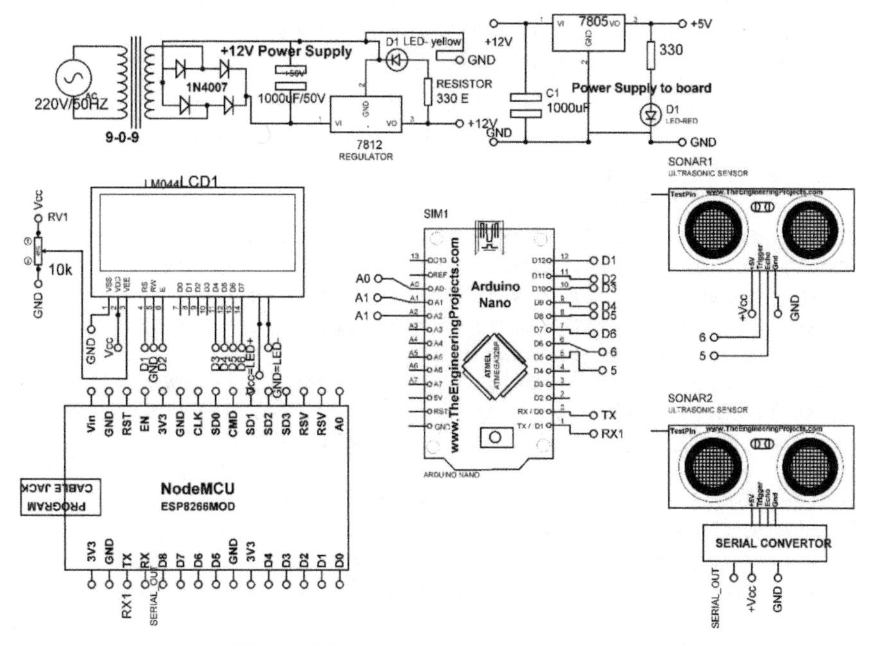

Figure 6.2 Circuit diagram of the system.

6.3 Program Code

6.3.1 Program Code for Arduino Nano

```
#include <LiquidCrystal.h>
const int RS = 12, EN = 11, D4 = 10, D5 = 9, D6 = 8, D7 = 7;
LiquidCrystal lcd(RS, EN, D4, D5, D6, D7);
String inputString_ULTRA = "";
String  ULTRA_SERIAL;
const int trigPin = 6;// connect trigger pin to pin 6
const int echoPin = 5;// connect trigger pin to pin 6
long duration_TRIG_ECHO;
int distance_TRIG_ECHO;
void setup()
{
 Serial.begin(9600); // initialize serial communication
 lcd.print(20,4);// initialize LCD
 pinMode(trigPin, OUTPUT); // set trigger pin as output
 pinMode(echoPin, INPUT); // set echo pin as input
 lcd.setCursor(0,0);// set the cursor on LCD
 lcd.print("Coolant and Oil");// print string on LCD
 lcd.setCursor(0,1);// set the cursor of LCD
```

```
 lcd.print("level monitoring");// print string on LCD
 delay(100);// wait for 100mSec
}


void loop()
{
 ULTRASONIC_READ();// function to read ultrasonic sensor
 digitalWrite(trigPin, LOW);// make trigger pin LOW
 delayMicroseconds(2);// wait for 10 micro second
 digitalWrite(trigPin, HIGH);// make trigger pin HIGH
 delayMicroseconds(10);// wait for 10 micro second
 digitalWrite(trigPin, LOW);// make trigger pin LOW
 duration_TRIG_ECHO = pulseIn(echoPin, HIGH);// read pulse
 distance_TRIG_ECHO = (duration/2) / 29.1;// calculate
    distance
 lcd.setCursor(0,2);// set cursor on LCD
 lcd.print("DIS1_SERIAL:");// print string on LCD
 lcd.print(ULTRA_SERIAL);// print integer on LCD
 lcd.setCursor(0,3);// set cursor on LCD
 lcd.print("DIS2_SERIAL:");// print string on LCD
 lcd.print(distance_TRIG_ECHO);// print variable on LCD
 Serial.print(ULTRA_SERIAL);// print serial the value
 Serial.print(";");// print string on serial
 Serial.print(distance_TRIG_ECHO); //print serial the value
 Serial.print('\n');// print new line char on serial
 delay(5000);// wait for 5 Sec
}


void ULTRASONIC_READ()// function to read serial data
{
 while (Serial.available()>0)// check serial data
 {
inputString_ULTRA = Serial.readStringUntil('\r');// Get
    serial input from ultrasonic sensor
ULTRA_SERIAL=String(((inputString_ULTRA[0]-48)*100) +
((inputString_ULTRA[1]-48)*10)+((inputString_ULTRA[2]-48)*1))
    +"."+String(((inputString_ULTRA[4]-48)*10)+((
    inputString_ULTRA[5]-48)*1));
 }
  inputString_ULTRA  = "";// make empty the string
  delay(20);// wait for 20mSec
}
```

6.3.2 Program Code of NodeMCU for ThingSpeak Server

```
#include <ESP8266WiFi.h>// add header for ESP8266
#include "StringSplitter.h"// add header for string splitter
String apiKey1 = "O44YTW0Z5WNO17N8";// add API key here
const char* ssid = "ESPServer_RAJ";// ID of Hot  spot
const char* password = "RAJ@12345";// password for hotspot
const char* server = "api.thingspeak.com";
WiFiClient client;
String OIL_LEVEL,COOLANT_LEVEL:// define string
String inputString_NODEMCU = "";   // a string to hold
    incoming data


        void setup()
        {
        Serial.begin(115200);// initialize serial
    communication
        delay(10);// wait for 10 sec
        WiFi.begin(ssid, password);// initialize wi-fi
    communication
        Serial.println();
        Serial.println();
        Serial.print("Connecting to ");// print serial data
        Serial.println(ssid);// print ssid


        while (WiFi.status() != WL_CONNECTED)
        {
        delay(500);// wait for 500 mSec
        Serial.print(".");// print string on LCD
        }
        Serial.println("");// print serial
        Serial.println("WiFi connected");// print string on
    LCD
        }


         void loop()
         {

              if (client.connect(server,80)) // check client
               {
               ULTRASONIC_serialEvent_NODEMCU();// call
    serial event
```

```
            send1_TX_OIL_COOLANT_PARA();// call the
    function to write data to thingspeak
           }
            client.stop();// command to stop client
            Serial.println("Waiting");// serial print the
    string
            delay(20000);// thingspeak needs minimum 15
    sec delay between updates

      }

void send1_TX_OIL_COOLANT_PARA()//call the function to write
    data to thingspeak
{
    String postStr = apiKey1;
    postStr +="\&field1=";
    postStr += String(OIL_LEVEL);
    postStr +="\&field2=";
    postStr += String(COOLANT_LEVEL);
    postStr += "\r\n\r\n";


    client.print("POST /update HTTP/1.1\n");
    client.print("Host: api.thingspeak.com\n");
    client.print("Connection: close\n");
    client.print("X-THINGSPEAKAPIKEY: "+apiKey1+"\n");
    client.print("Content-Type: application/x-www-form-
    urlencoded\n");
    client.print("Content-Length: ");
    client.print(postStr.length());
    client.print("\n\n");
    client.print(postStr);
    Serial.print("Send data to channel-1 ");// print string
    on serial port
    Serial.print("Content-Length: ");// print string on
    serial port
    Serial.print(postStr.length());// print string on serial
    port
    Serial.print("Field-4: ");// print string on serial port
    Serial.print(OIL_LEVEL);// print value on serial port
    Serial.print("Field-5: ");// print string on serial port
    Serial.print(COOLANT_LEVEL);// print value on serial port
    Serial.println(" data send");// print string on LCD
}
void ULTRASONIC_serialEvent_NODEMCU() // function to read
    serial data from
```

```
Arduino Uno

{
  while (Serial.available()>0)// check data on serial RX pin
 {
inputString_NODEMCU = Serial.readStringUntil('\n');// Get
    serial input
StringSplitter *splitter = new StringSplitter(
    inputString_NODEMCU, ',', 3);   // new
StringSplitter(string_to_split, delimiter, limit)
int itemCount = splitter->getItemCount();
 for(int i = 0; i < itemCount; i++)
  {
   String item = splitter->getItemAtIndex(i);
   OIL_LEVEL= splitter->getItemAtIndex(0);// data of oil
    level
   COOLANT_LEVEL = splitter->getItemAtIndex(1);// data of
    coolant level
  }
  inputString_NODEMCU = "";// make string empty
  delay(200);// wait for 200mSec
  }
}
```

6.4 ThingSpeak Server

Follow the steps discussed in Section 5.4 and check the sensory data on the server. Figure 6.3 shows the oil level and Figure 6.4 shows the coolant level w.r.t time

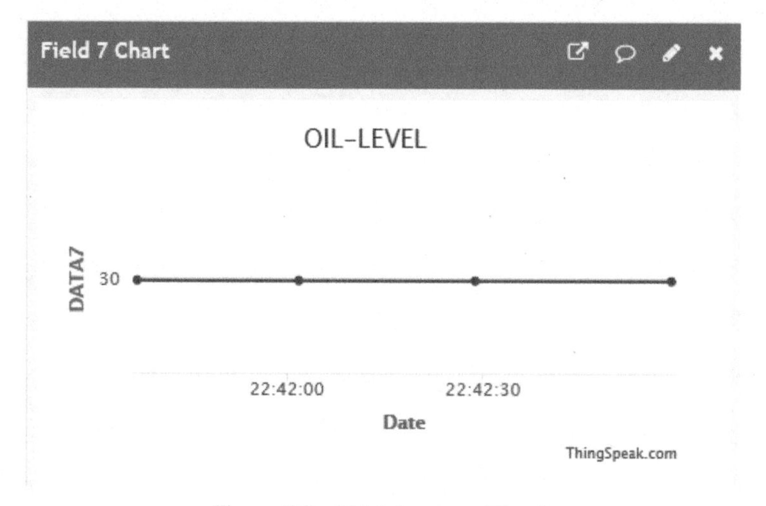

Figure 6.3 Field showing oil level.

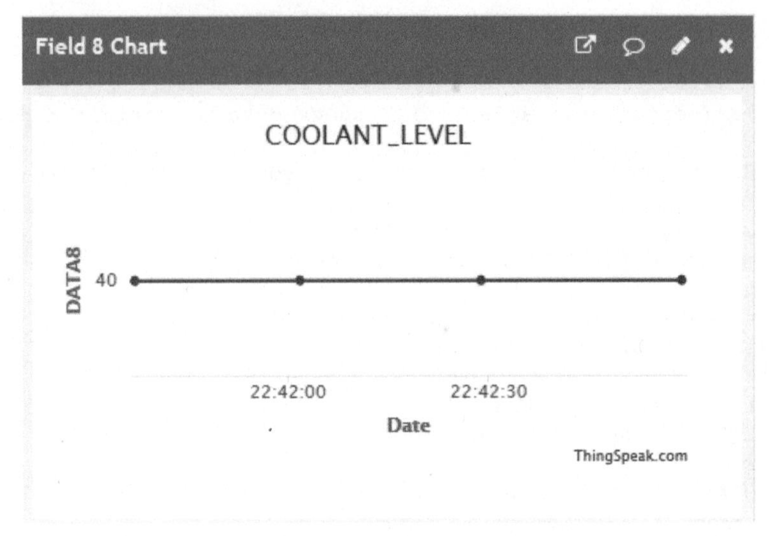

Figure 6.4 Field showing coolant level.

7

Fleet and Driver Management System

In order to trace the vehicle position and optimum utilization of fleet, and further to monitor the cabin atmosphere like weather driver is alcoholic or smoking while driving the fleet and driver management system is important. In this chapter an IoT based system is proposed.

7.1 Introduction

Figure 7.1 shows the block diagram of the system with Ti Launch PAD, NodeMCU, and other devices. It comprises a +12 V/500 mA power supply, a 12 to 5 V convertor, a Ti Launch PAD, a NodeMCU, an LCD, a GPS, a smoke sensor, and an alcohol sensor. The main objective is to display the data of fleet and driver management system using the smoke sensor, alcohol sensor, and GPS coordinates on the LCD by reading the sensors and GPS values. The sensory information from the Ti Launch PAD is transferred to the NodeMCU via a serial link. The information with time is uploaded to the cloud server using the NodeMCU/WiFi modem. The information is also communicated to the mobile App. Table 7.1 shows the list of the components to develop the system.

7.2 Circuit Diagram

The following are the interfacing connections for the system.

1. +5V pin of the power supply is connected to the Vcc pin of the launch pad and the NodeMCU.
2. GND pin of the power supply is connected to the GND pin of the Ti Launch pad and the NodeMCU.
3. Pins 1 and 16 of the LCD are connected to GND of the power supply.
4. Pins 2 and 15 of the LCD are connected to +Vcc of the power supply.

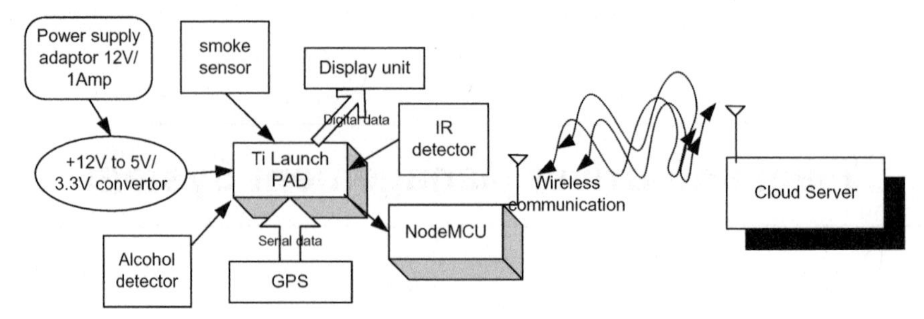

Figure 7.1 Block diagram of the system.

Table 7.1 Components' list

Name of Components	Quantity
NodeMCU	1
Ti Launch Pad	1
LCD20*4	1
LCD20*4 patch	1
DC 12V/1Amp adaptor	1
12 to 5 V, 3.3 V converter	1
LED with 330 Ω resistor	1
Jumper wire M to M	20
Jumper wire M to F	20
Jumper wire F to F	20
Smoke sensor MQ6	1
Alcohol sensor MQ3	1
GPS	1
IR detector	1

5. Two fixed lags of POT are connected to +5V and GND of the LCD and variable lag of the POT is connected to pin 3 of the LCD, respectively.
6. RS, RW, and E pins of the LCD are connected to pin D1, GND, and D2 of the Ti Launch PAD, respectively.
7. D4, D5, D6, and D7 pins of the LCD are connected to pins D3, D4, D5, and D6 of the Ti Launch PAD..
8. +5V and GND pins of the **MQ3 sensor** are connected to +5V and GND pins of the power supply, respectively.
9. OUT1 of the **MQ3 sensor** is connected to A4 of the Ti Launch PAD.
10. +5V and GND pins of the **MQ6 sensor** are connected to +5V and GND pins of the power supply, respectively.
11. OUT2 of the **MQ6 sensor** is connected to A5 of the Ti Launch PAD.

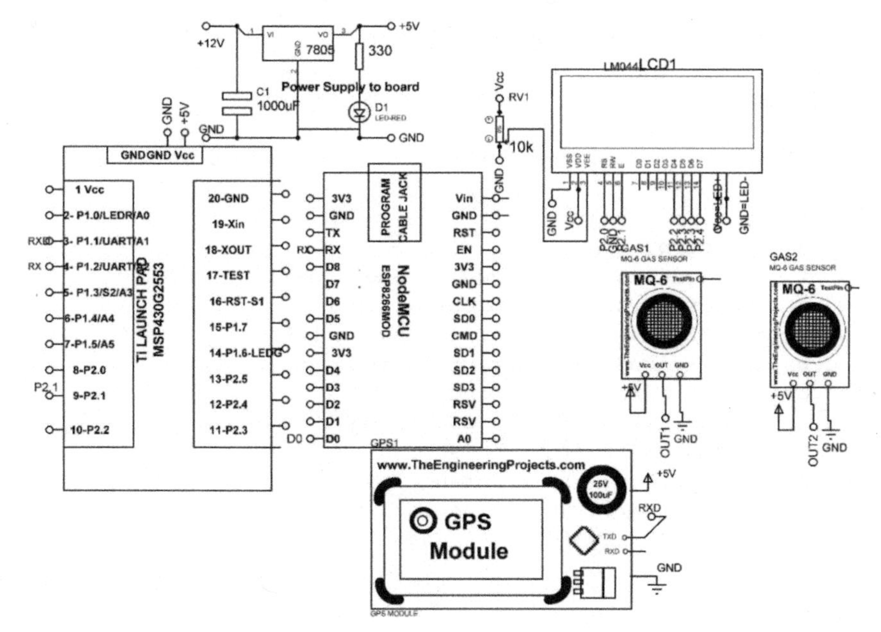

Figure 7.2 Circuit diagram of the system.

12. Connect +Vcc, GND, and OUT pins of the GPS to +5V, GND, and RX pins of the launch pad.
13. Connect the P1.2 pin of the Ti Launch PAD to the RX pin of the NodeMCU.

Figure 7.2 shows the circuit diagram of the system.

7.3 Program Code

7.3.1 Program Code for Ti Launch Pad with Energeia IDE

```
/////////////// library for LCD
#include <LiquidCrystal.h>
LiquidCrystal lcd(P2_0, P2_3, P2_4, P2_5, P2_6, P2_7);//
    define pin of LCD in Ti Launch pad
//// library for GPS
#include <TinyGPS.h>
TinyGPS gps;
void getgps(TinyGPS &gps);
float latitude, longitude;
byte a;
```

```
////////////// for Ultrasonic sensor
String inputString_ULTRA = "";         // a string to hold
   incoming data
String ULTRA;
void getgps(TinyGPS &gps)
{
 float latitude, longitude;
 decode and display position data
 gps.f_get_position (&latitude, &longitude);
 lcd.setCursor (0,3); // set cursor on LCD
 lcd.print ("Lat:"); // print string on LCD
 lcd.print (latitude,5);//print latitude value on LCD
 lcd.print(" ");// print string on LCD
 lcd.setCursor(10,3); // set cursor on LCD
 lcd.print("Long:"); // print string on LCD
 lcd.print(longitude,5); // print longitude value on LCD
 lcd.print(" ");// print string on LCD
 delay(3000); // wait for 3 seconds
}


void  CALL_GPS()
{
 byte a;
 if (Serial.available() > 0) // if there is data coming into
   the serial line
 {
  a = Serial.read(); // get the byte of data
   if(gps.encode(a)) // if there is valid GPS data...
     {
         getgps(gps); // grab the data of GPS
     }
 }


 }
void setup()
{
 lcd.begin(20, 4); // initialize LCD
 Serial.begin(9600); // initialize serial communication
 lcd.setCursor(0,0); // set cursor on LCD
 lcd.print("Fleet tracking"); // print string on LCD
 lcd.setCursor(0,1); // set cursor on LCD
 lcd.print("and monitoring sys");// set cursor on LCD
 delay(4000); // wait for 40 sec
 lcd.clear();// clear the contents of LCD
```

```
}

void loop()
{
int MQ3=analogRead(A0); // read analog pin A0 of MQ3 sensor
int MQ6=analogRead(A1); // read analog pin A1 of MQ6 sensor
////////////////////////// read and display Water flow
 lcd.setCursor(0,1);// set the cursor on LCD
 lcd.print("Alcohal_Level:");// print string on LCD
 lcd.print(MQ3); // print the levels of MQ3 sensor on LCD
 lcd.setCursor(0,2);  // set the cursor on LCD
 lcd.print("Smoke_Level:"); // print string on LCD
 lcd.print(MQ6); // print the levels of MQ3 sensor on LCD
 Serial.print(MQ3); // print the levels of MQ3 sensor on
    serial port
 Serial.print(",");// print string on LCD
 Serial.print(MQ6); // print the levels of MQ6 sensor on
    serial port
 Serial.print(","); // print string on LCD
 Serial.print(latitude); // print the value of latitude on
    serial port
 Serial.print(",");// print string on serial port
 Serial.print(longitude); // print the value of latitude on
    serial port
 Serial.print('\n'); // print string on serial port
}
```

7.3.2 Program Code for NodeMCU with Arduino IDE

```
#define CAYENNE_PRINT Serial
#include <CayenneMQTTESP8266.h>
#include <ESP8266WiFi.h>
#include "StringSplitter.h"


// add the credentials of Wi-Fi
char ssid[] = "ESPServer_RAJ"; // hotspot ID
char wifiPassword[] = "RAJ@12345";// hotspot password


// Cayenne authentication info. This should be obtained from
    the Cayenne Dashboard.
char username[] = "fac81bb0-7283-11e7-85a3-9540e9f7b5aa";
char password[] = "3745eb389f4e035711428158f7cdc1adc0475946";
char clientID[] = "386b86f0-7284-11e7-b0bc-87cd67a1f8c7";
```

```
String inputString_NODEMCU = "";// take string to store
    serial data
int ALCOHAL,SMOKE,latitude,longitude;// define integer
void setup()
{
 pinMode(D0, OUTPUT); // make D0 pin as output
 Serial.begin(9600); // initialize serial communication
 Cayenne.begin(username, password, clientID, ssid,
    wifiPassword);
}

void loop()
{
Cayenne.loop();// initialize cayenne loop
SerialDATA(); // call function for serial data recording
Cayenne.virtualWrite(0, SMOKE); // write variable value on
    channel 0 of Cayenne dashboard
Cayenne.virtualWrite(1, ALCOHAL);// // write variable value
    on
channel 0 of Cayenne dashboard
Cayenne.virtualWrite(2, latitude);// // write variable value
    on
channel 0 of Cayenne dashboard
Cayenne.virtualWrite(3, longitude);// // write variable value
     on channel 0 of Cayenne dashboard
 delay(500);// wait for 500 mSec
}

CAYENNE_IN_DEFAULT()// cayenne function to write input from
    dashboard to hardware
{
CAYENNE_LOG("CAYENNE_IN_(1)(%u) - %s, %s", request.channel,
    getValue.getId(), getValue.asString());
int i = getValue.asInt();
if (i>=45)
   {
   digitalWrite(D0,HIGH); // Make D0 pin HIGH
   }
   else
   {
   digitalWrite(D0,LOW); // Make D0 pin HIGH
   }

}
```

```
void serialEvent_NODEMCU() // function to read serial data
    from Ti launch pad
{
  while (Serial.available()>0)// check serial data
  {
   inputString_NODEMCU = Serial.readStringUntil('\n');//
    Get serial input and store in string


StringSplitter *splitter = new StringSplitter
    (inputString_NODEMCU, ',', 4);  // new
StringSplitter(string_to_split, delimiter, limit)
  int itemCount = splitter->getItemCount();
  for(int i = 0; i < itemCount; i++)
  {
   String item = splitter->getItemAtIndex(i);
   SMOKE = splitter->getItemAtIndex(0); // split and record
    smoke sensor data
   ALCOHAL= splitter->getItemAtIndex(1); // split and record
    alcohol sensor data
   latitude= splitter->getItemAtIndex(2); // split and record
     latitude data
   logitude=splitter->getItemAtIndex(3); // split and record
    lonitude data
  }
  inputString_NODEMCU = ""; // make string empty
  delay(200); // wait for 200 mSec
  }
}
```

7.4 Cayenne APP

Steps to Add NodeMCU in Cayenne Cloud

1. Install the Arduino IDE and add Cayenne MQTT Library to Arduino IDE.
2. Install the ESP8266 board package to Arduino IDE.
3. Install the required USB driver on the computer to program the ESP8266.
4. Connect the ESP8266 to PC/Mac via the data-capable USB cable.
5. In the Arduino IDE, go to the **tools** menu, select the **board**, and now the **port** ESP8266 is connected to.
6. Use the MQTT username, MQTT password, client ID, as well as ssid[] and wifiPassord[] in the arduino IDE to write code (Figure 7.3).

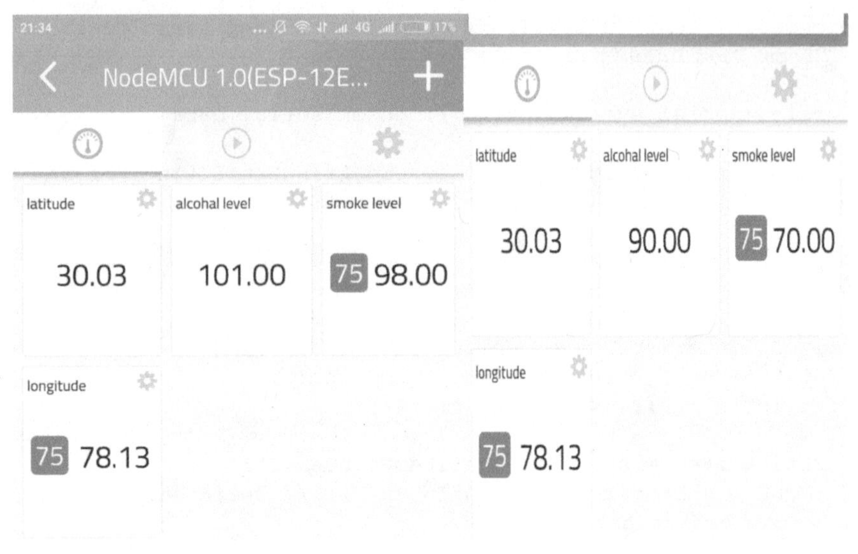

Figure 7.3 Cayenne APP.

8

Smart Road Communication System for Mobile Vehicles

In order to avoid the anomalies of roads and bridges, different sensors need to be installed. These sensors may help to predict the natural disasters and communicate in advance to the traffic and its controlling departments. In the real scenario, there are black zones on the road where no internet signals are available. To develop the system, two different nodes need to be designed: one node for black zone and other for communicating to the cloud through the Internet.

8.1 Introduction

Figure 8.1 shows the block diagram of the road unit for black zone with Ti Launch PAD, XBee (to communicate within black zone), and other devices. It comprises a +12 V/500 mA power supply, a 12 to 5 V convertor, a Ti Launch PAD, an XBee, an LCD, a GPS, a DHT11 sensor, BMP180, and a rain sensor. The main objective is to display the sensory data for the smart road management system using the DHT11 sensor, BMP180, and rain sensor on the LCD by reading the sensors. The sensory information from the Ti Launch PAD is transferred to the XBee via a serial link. This information is communicated to other nodes to reach at node which has IoT facility.

Figure 8.2 shows the block diagram of the unit at the mobile vehicle. It comprises a +12 V/500 mA power supply, a 12 to 5V convertor, a Ti Launch PAD, an XBee, and an LCD. The main objective is to receive the information through the XBee modem and display the data using the LCD. Then the sensory information from the Ti Launch PAD is transferred to the GPRS modem via a serial link. The information is uploaded to the cloud server ThingSpeak using the GPRS modem.

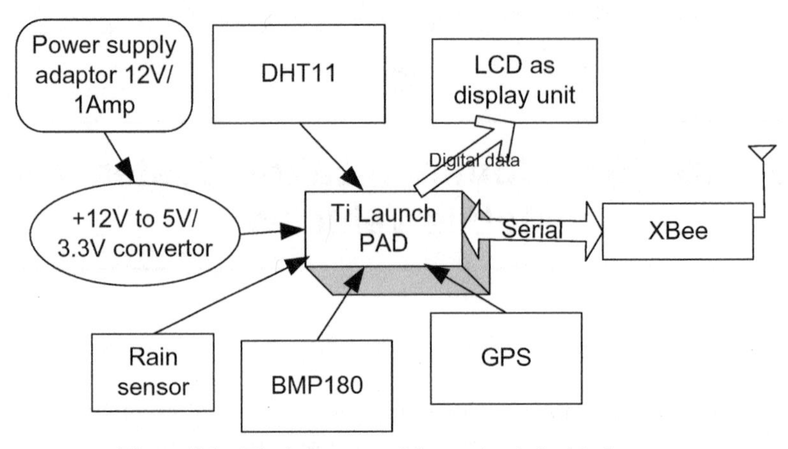

Figure 8.1 Block diagram of the road unit for black zone.

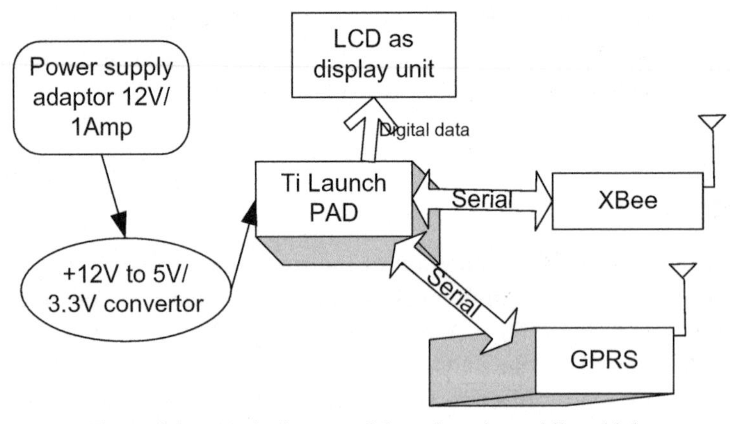

Figure 8.2 Block diagram of the unit at the mobile vehicle.

Tables 8.1 and 8.2 show the list of the components required to design a system.

8.2 Circuit Diagram

8.2.1 Circuit Diagram for the Road Unit for Black Zone

1. GND pin of the power supply is connected to the GND pin of the launch pad and the Ti Launch PAD.
2. Pins 1 and 16 of the LCD are connected to GND of the power supply.
3. Pins 2 and 15 of the LCD are connected to +Vcc of the power supply.

Table 8.1 Components' list for of the road unit for black zone

S.No	Components	Quantity
1	GPS	1
2	LCD20*4	1
3	LCD20*4 patch	1
4	DC 12 V/1 A adaptor	1
5	12 to 5 V, 3.3 V converter	1
6	LED with 330 Ω resistor	1
7	Jumper wire M to M	20
8	Jumper wire M to F	20
9	Jumper wire F to F	20
10	DHT11	1
11	BMP180	1
12	Rain sensor	1
13	Ti Launch PAD	1
14	XBee	1
15	XBee breakout board	1

Table 8.2 Components' list for of the unit at the mobile vehicle

S.No	Components	Quantity
1	GPRS modem	1
2	LCD20*4	1
3	LCD20*4 patch	1
4	DC 12 V/1 A adaptor	1
5	12 to 5 V, 3.3 V converter	1
6	LED with 330 Ω resistor	1
7	Jumper wire M to M	20
8	Jumper wire M to F	20
9	Jumper wire F to F	20
10	Ti Launch PAD	1
11	XBee	1
12	XBee breakout board	1

4. Two fixed lags of the POT are connected to +5V and GND of the LCD and variable lag of the POT is connected to pin 3 of the LCD.
5. RS, RW, and E pins of the LCD are connected to pins D1 = P2.0, GND, and D2 = P2.1 of the Ti Launch PAD.
6. D4, D5, D6, and D7 pins of the LCD are connected to pins D3 = P2.2, D4 = P2.3, D5 = P2.4, and D6 = P2.5 of the Ti Launch PAD.
7. +5V and GND pins of the **DHT sensor** are connected to +5V and GND pins of the power supply, respectively.
8. OUT1 of the **DHT sensor** is connected to P1_3 of the Ti Launch PAD.

9. +5V and GND pins of the **BMP180 sensor** are connected to +5V and GND pins of the power supply, respectively.
10. SCL and SDA lines of the **BMP sensor** are connected to A4 and A5 of the Ti Launch PAD.
11. +5V and GND pins of the **rain sensor** are connected to +5V and GND pins of the power supply, respectively.
12. OUT1 of the **RAIN sensor** is connected to P1_0 of the Ti Launch PAD.
13. Connect +Vcc, GND, and OUT pins of the **GPS** to +5V, GND, and RX pins of the launch pad.
14. Connect the P1.2 pin of the Ti Launch PAD to the RX pin of the NodeMCU.
15. Connect +Vcc, GND, and TX pins of the XBee to +5V, GND, and P1_1 pins of the Ti Launch PAD. Figure 8.3 shows the circuit diagram of the road unit for black zone.

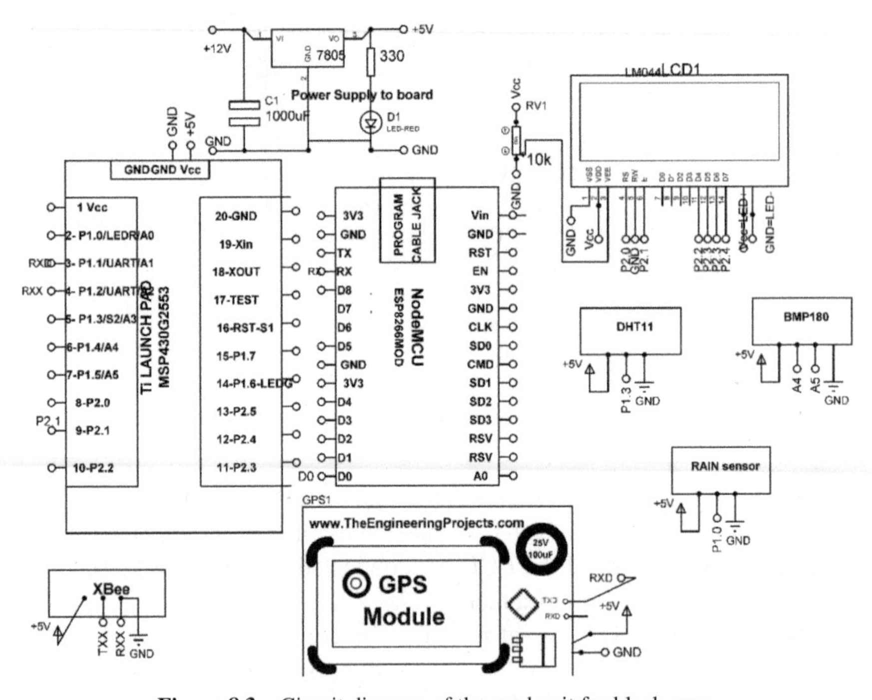

Figure 8.3 Circuit diagram of the road unit for black zone.

8.2.2 Circuit Diagram of the Unit at the Mobile Vehicle

1. +5V pin of the power supply is connected to the Vcc pin of the launch pad and GPRS modem.
2. GND pin of the power supply is connected to the GND pin of the launch pad and the Ti Launch Pad.
3. Pins 1 and 16 of the LCD are connected to GND of the power supply.
4. Pins 2 and 15 of the LCD are connected to +Vcc of the power supply.
5. Two fixed lags of the POT are connected to +5V and GND of the LCD and variable lag of the POT is connected to pin 3 of the LCD.
6. RS, RW, and E pins of the LCD are connected to pins D1, GND, and D2 of the Ti Launch PAD.
7. D4, D5, D6, and D7 pins of the LCD are connected to pins D3, D4, D5, and D6 of the Ti Launch PAD.
8. Connect +Vcc, GND, and OUT pins of the GPS to +5V, GND, and RX pins of the launch pad.
9. Connect the P1.2 pin of the Ti Launch PAD to the RX pin of the GPRS modem.
10. Connect +Vcc, GND, and RX pins of the XBee to +5V, GND, and P1.2 pins of the Ti Launch PAD. Figure 8.4 shows the circuit diagram of the unit at the mobile vehicle.

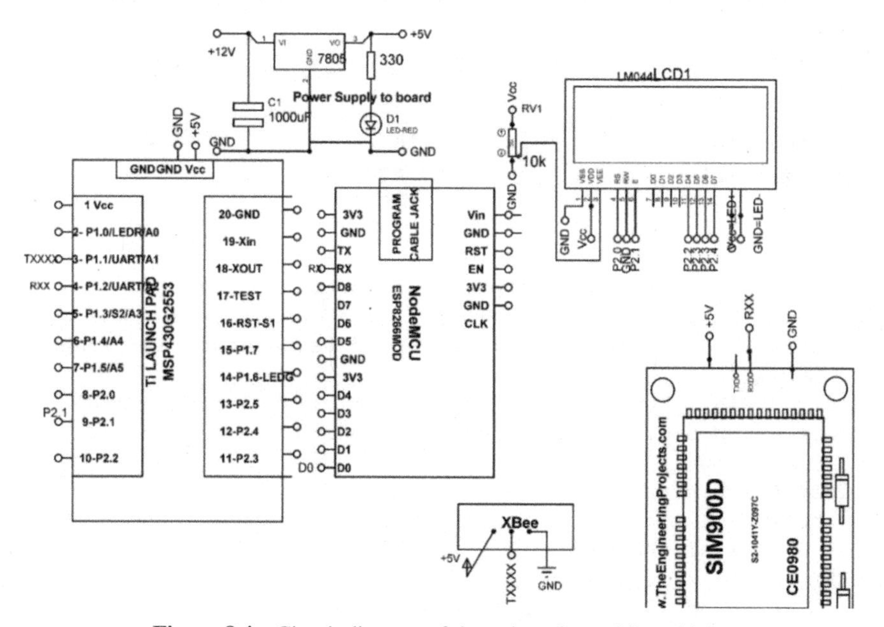

Figure 8.4 Circuit diagram of the unit at the mobile vehicle.

8.3 Program Code from Arduino IDE

8.3.1 Program Code for Ti Launch PAD with Energeia IDE

```
///// library for BMP185
#include <Wire.h> // library for I2C mode
#include <Adafruit_BMP085.h> // library for BMP185 sensor
Adafruit_BMP085 bmp;


//// library for GPS
#include <TinyGPS.h> // library for GPS
TinyGPS gps;
void getgps(TinyGPS &gps);
float latitude, longitude;// assign float
byte a; // assume Byte
//////// library for DHT11
#include <dht.h>
dht DHT; // define
#define DHT11_PIN 2 // define pin of DHT11 sensor


/////////////// library for LCD
#include <LiquidCrystal.h> // attach library for LCD
LiquidCrystal lcd(13, 12, 11, 10, 9, 8);// pin of Arduino
    Nano is connect to LCD
void setup()
{
 Serial.begin(9600); // initialize serial communication
 lcd.begin(20, 4); // // initialize LCD
 bmp.begin();// initialize BMP180 sensor
}
void loop()
{
 lcd.clear(); // clear the previous contents from LCD
int chk = DHT.read11(DHT11_PIN); // read DHT11 sensor
 float TEMP=DHT.temperature; // record temperature as float
    value
 float HUM=DHT.humidity; // record humidity as float value
 float PRESS=bmp.readPressure(); // record pressure value as
    float
 float ALT=bmp.readAltitude(); // record altitude value as
    float
 int RAIN=analogRead(A0); // read analog pin A0
 CALL_GPS(); // call GPS
  lcd.setCursor(0,0); // set cursor on LCD
```

```
 lcd.print("TEMP:"); // print string on LCD
 lcd.print(TEMP); // print value of temperature on LCD
 lcd.setCursor(10,0); // set cursor on LCD
 lcd.print("HUM:"); // print string on LCD
 lcd.print(HUM); // print value of humidity on LCD


//////////////////////////// read and display BMP185 data
 lcd.setCursor(0,1); // set cursor on LCD
 lcd.print("P0:"); // print string on LCD
 lcd.print(PRESS); // print value of pressure on LCD
 lcd.print("Pa"); // print string on LCD


 lcd.setCursor(10,1);  // Calculate altitude assuming '
   standard' barometric & pressure of 1013.25 millibar =
   101325 Pascal
 lcd.print("A0:"); // print string on LCD
 lcd.print(ALT); // print value of altitude on LCD
 lcd.print("m"); // print string on LCD


 lcd.setCursor(0,2);  // Calculate altitude assuming '
   standard' barometric & pressure of 1013.25 millibar =
   101325 Pascal
 lcd.print("RAIN_LEVEL:"); // print string on LCD
 lcd.print(RAIN); // print value of rain sensor on LCD


 Serial.print(TEMP); // print value of temperature on serial
    port
 Serial.print(","); // print string on serial port
 Serial.print(HUM); // print value of humidity on serial
   port
 Serial.print(","); // print string on serial port
 Serial.print(PRESS); // print value of pressure on serial
   port
 Serial.print(","); // print string on serial port
 Serial.print(ALT); // print value of altitude on serial
   port
 Serial.print(","); // print string on serial port
 Serial.print(RAIN); // print value of rain on serial port
 Serial.print(","); // print string on serial port
 Serial.print(latitude); // print value of latitude on
   serial port
 Serial.print(","); // print string on serial port
```

```
  Serial.print(longitude); // print value of longitude on
    serial port
  Serial.print('\n'); // print string on serial port
  delay(30);// wait for delay of 30 mSec

}


void getgps(TinyGPS &gps) // function to get the coordinate
    of GPS
{
 float latitude, longitude;
 decode and display position data
 gps.f_get_position(&latitude, &longitude);
 lcd.setCursor(0,3); // set cursor on LCD
 lcd.print("Lat:"); // print string on LCD
 lcd.print(latitude,5); // print value of latitude on LCD
 lcd.print(" ");// print string

 lcd.setCursor(10,3); // set cursor on LCD

 lcd.print("Long:"); // print string on LCD
 lcd.print(longitude,5); // print value of lonitude on LCD
 lcd.print(" "); // print string on LCD
 delay(3000); // wait for 3 seconds
}


void  CALL_GPS()
{
 byte a;
 if (Serial.available() > 0 ) // if there is data coming into
     the serial line
 {
  a = Serial.read(); // record the serial value in variable
   if(gps.encode(a)) // if there is valid GPS data...
   {
    getgps(gps); // get the data and display it on the LCD
    }
 }

 }
```

8.3.2 Program Code for Ti Launch PAD with Energeia IDE and GPRS

```
#include <SoftwareSerial.h>// add softserial
library
#include <String.h> // add atring as library
SoftwareSerial MyGPRS (P1_6, P1_7); // make pins as
RX and TX pins
char thingSpeakAddress[] = "api.thingspeak.com";
//////////////// library for LCD
#include <LiquidCrystal.h>
LiquidCrystal lcd(P2_0, P2_1,P2_2, P2_3, P2_4, P2_5);
// LCD pin for Ti launch pad
//int8_t answer;
float answer;
float TEMP,HUM,PRESS,ALT,latitude,longitude; //
assume float
String inputString_NODEMCU = "";  // a string to hold
 incoming data


void CallGPRS()
{
gprspwr_on();
serialEvent_NODEMCU(); // call function to record
sensors data
//connect gprs to internet

answer = sendATcommand("AT+CGATT?","OK",5,2000); // sand AT
    command
answer = sendATcommand("AT+CSTT=\"CMNET\"","OK",3,2000); //
    send AT    command
answer = sendATcommand("AT+CIICR","OK",3,2000); // send AT
    command
answer = sendATcommand("AT+CIFSR","OK",3,2000); // send AT
    command
answer = sendATcommand("AT+CIPSPRT=0","OK",3,2000); // //send
     AT  command
//connect gprs to thingspeak
answer= sendATcommand("AT+CIPSTART=\"tcp\",\"api.thingspeak.
    com\",\"80\""," CONNECT OK", 5,2000);
serialEvent_NODEMCU(); // call function
            //post data to thingspeak
            int param1=TEMP;
            int param2=HUM;
            int param3=PRESS;
```

```
int param4=ALT;
int param5=RAIN;
int param6=latitude;
int param7=longitude;

  answer = senddata1(param1,param2,param3,param4,
  param5,param6,param7);
  delay(3000);
  gprspwr_off();


  //put arduino to sleep?
  for (int i=0; i<60; i++)
  {
    delay(150);
  }
  }
 /****************************************gprs
2nd function end***********************************/


void setup()
{
 // put your setup code here, to run once:
 MyGPRS.begin(9600);                 // the GPRS
 baud rate
 Serial.begin(9600);                    // the
 computer serial interface baud rate
 lcd.begin(20, 4); // initialize LCD
 delay(1000); // wait for 1000 mSec
 lcd.print("GPRS BASED IoT"); // print string on
 LCD
 delay(1000); // wait for 1000 mSec
}
void loop()
{
 byte l;
 serialEvent_NODEMCU(); // call function
 CallGPRS();   // call GPRS function
 delay(500); // wait for 500 mSec
}


/*****************************************/
int8_t senddata1(int data,int data1,int data2,int
data3,int data4,int data5,int data6,int data7)
```

```
        {

        MyGPRS.println("AT+CIPSEND");
        while(MyGPRS.available() > 0) MyGPRS.read();
// Clean the input buffer
        delay(500); // wait for 500 mSec
        MyGPRS.println("POST /update HTTP/1.1");    //
Send the AT command
        while(MyGPRS.available() > 0) MyGPRS.read();
// Clean the input buffer
        delay(500); // wait for 500 mSec
        MyGPRS.println("Host: api.thingspeak.com");
// Send the AT command
        while(MyGPRS.available() > 0) MyGPRS.read();
// Clean the input buffer
        delay(500); // wait for 500 mSec
        MyGPRS.println("Connection: close");    // Send
the AT command
        while(MyGPRS.available() > 0)MyGPRS.read();
// Clean the input buffer
        delay(500); // wait for 500 mSec
        MyGPRS.println("X-THINGSPEAKAPIKEY:
L5I8F6JM3NKUQNTU");//T1GIUPBKKRDPMWRX");
        while(Serial2.available() > 0) Serial2.read();
  // Clean the  input buffer
        delay(500);
        MyGPRS.println("Content-Type: application/x-www-
form-urlencoded");    //
        Send  the AT command


        while(MyGPRS.available() > 0) MyGPRS.read();
// Clean the input buffer
        delay(500);
        MyGPRS.println("Content-Length:92");    // Send
the AT command


        while(MyGPRS.available() > 0) MyGPRS.read();
// Clean the input buffer
        delay(500);
        MyGPRS.println("");    // Send the AT command

        while(MyGPRS.available() > 0) MyGPRS.read();
// Clean the input buffer
        delay(500);
```

```
MyGPRS.print("&field1=");     // Send the AT
command
MyGPRS.print(data);
MyGPRS.print("&field2=");     // Send the AT
command
MyGPRS.print(data1);
MyGPRS.print("&field3=");     // Send the AT
command
MyGPRS.print(data2);
MyGPRS.print("&field4=");     // Send the AT
command
MyGPRS.print(data3);
MyGPRS.print("&field5=");     // Send the AT
command
MyGPRS.print(data4);
MyGPRS.print("&field6=");     // Send the AT
command
MyGPRS.print(data5);
MyGPRS.print("&field7=");     // Send the AT
command
MyGPRS.print(data6);
MyGPRS.print("&field8=");     // Send the AT
command
MyGPRS.print(data7);
while(MyGPRS.available() > 0) MyGPRS.read();
// Clean the input buffer
delay(500); // wait for 500 mSec
MyGPRS.println((char)26);
delay(500); // wait for 500 mSec
while(MyGPRS.available() > 0) MyGPRS.read();
// Clean the input buffer
delay(500); // wait for 500 mSec


answer = 0;
return answer;
}


void gprspwr_on()
{
 pinMode(5, OUTPUT);  // make pin 5 as output
 pin
 digitalWrite(5,LOW); // make pin 5 as low
 delay(1000); // wait for 1000 mSec
```

```
 digitalWrite(5,HIGH); // make pin 5 as high
 delay(2000); // wait for 1000 mSec
 digitalWrite(5,LOW); // make pin 5 as low
 readATcommand("Call Ready",6,10000);
 if (answer == 1)
 {
 }
}

void gprspwr_off()
{
pinMode(5, OUTPUT);  //set pin5 as output
digitalWrite(5,LOW); // make pin5 as LOW
delay(1000); // wait for 1000 mSec
digitalWrite(5,HIGH); // make pin5 as LOW
delay(2000); // wait for 1000 mSec
digitalWrite(5,LOW); // make pin5 as LOW
answer = readATcommand("NORMAL POWER DOWN",
2,2000);
 if (answer == 1)
 {
 }
}

boolean gprspwr_status()
{
 answer = sendATcommand("AT", "OK", 2, 2000);
 if (answer == 0)
{


}
 else if (answer == 1)
{

}
 return answer;
}
int8_t readATcommand(char* expected_answer1,
unsigned int expected_answers, unsigned
int timeout)
{
 uint8_t x=0,  answer=0;
 boolean complete = 0;
 char a;
 char response[100];
```

```
unsigned long previous;
String  incomingdata;
boolean first;
previous = millis();
for(int i = 0; i < expected_answers; i++)
{
 x = 0;
 complete = 0;
 a = 0;
 first = 0;
 memset(response, '\0', 100);     // Initialize
 the string do
 {
  if(MyGPRS.available() != 0)
  {
   a = MyGPRS.read();
   //Serial.println(a,DEC);
   if (a == 13)
   {
    a = MyGPRS.read();
    if (a == 10)
    {
     if (first == 0)
     {
     }
     else
     {
      complete = 1;
     }
    }
   }
   else if(a == 0)
   {
   }
   else
   {
    response[x] = a;
    x++;
    first = 1;
   }
   if(strstr(response, expected_answer1)
   != NULL)
   {
    answer = 1;
    complete = 1;
    return answer;
```

```
    }
    else if(strstr(response, "ERROR") != NULL)
    {
     answer = 2;
    }
   }
  }
 while((complete == 0) && ((millis() - previous)
 < timeout));


 }


 return answer;

}


int8_t sendATcommand(char* ATcommand, char*
expected_answer1, unsigned
int expected_answers, unsigned int timeout)
{

 uint8_t x=0,  answer=0;
 boolean complete = 0, first = 0;
 char a;
 char response[100];
 unsigned long previous;
 String  incomingdata; // assign string
 delay(100); // wait for 100mSec
 while(MyGPRS.available() > 0) MyGPRS.read();
  // Clean the input buffer
 MyGPRS.println(ATcommand);    // Send the AT
 command
 previous = millis();
  for(int i = 0; i < expected_answers; i++){
  x = 0;
  complete = 0;
  a = 0;
  first = 0;
  memset(response, '\0', 100);    // Initialize
  the string
  do{
   if(MyGPRS.available() != 0)
   {
```

```
         a = MyGPRS.read();
         if (a == 13)
         {
          a = MyGPRS.read();
          if(a == 10){
            if (first == 0)
            {
             //keep going, just ignore it
            }
            else
            {
             complete = 1;
            }
           }
          }
          else if(a == 0)
          {

            }
          else
          {
           response[x] = a;
           x++;
           first = 1;

          }
          if (strstr(response, expected_answer1)
          != NULL)
          {
           answer = 1;
           complete = 1;
          }
          else if(strstr(response, "ERROR") != NULL)
          {
           answer = 2;
           complete = 1;
          }
         }
        }
        while((complete == 0) && ((millis() -
        previous) < timeout));


        }
        return answer;
```

```
}
void serialEvent_NODEMCU()
{
while (mySerial.available()>0)
 {
inputString_NODEMCU = mySerial.readStringUntil
('\n');// Get serial input
StringSplitter *splitter = new StringSplitter
(inputString_NODEMCU, ',', 8);  //
new StringSplitter(string_to_split, delimiter,
limit)
int itemCount = splitter->getItemCount();


 for(int i = 0; i < itemCount; i++)
  {
   String item = splitter->getItemAtIndex(i);
   TEMP = splitter->getItemAtIndex(0); //
   store temperature
   HUM= splitter->getItemAtIndex(1); // store
   humidity
   PRESS = splitter->getItemAtIndex(2); //
   store pressure
   ALT= splitter->getItemAtIndex(3); // store
   altitude
   RAIN= splitter->getItemAtIndex(4); // store
   rain level
   latitude= splitter->getItemAtIndex(5); //
   store latitude
   logitude=splitter->getItemAtIndex(6); //
   store lonitude
  }
  inputString_NODEMCU = ""; // make string
  empty
  delay(200); // wait for 200 mSec
  }
```

8.4 ThingSpeak Server

Follow the steps discussed in Section 5.4 and check the sensory data on different fields of the server. Figure 8.5 shows the temperature sensor reading, Figure 8.6 shows the humidity sensor reading, Figure 8.7 shows the air pressure, Figure 8.8 shows the altitude, Figure 8.9 shows the latitude and Figure 8.10 shows the longitude.

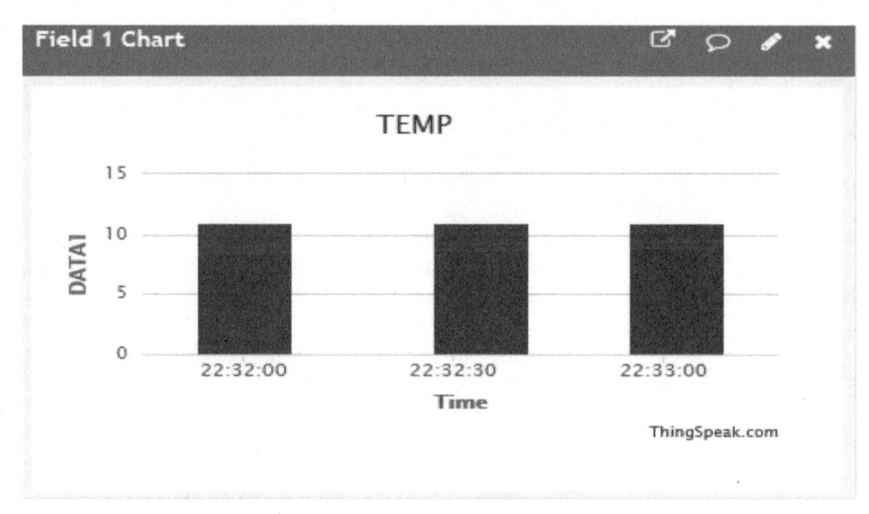

Figure 8.5 Field showing temperature sensor readings.

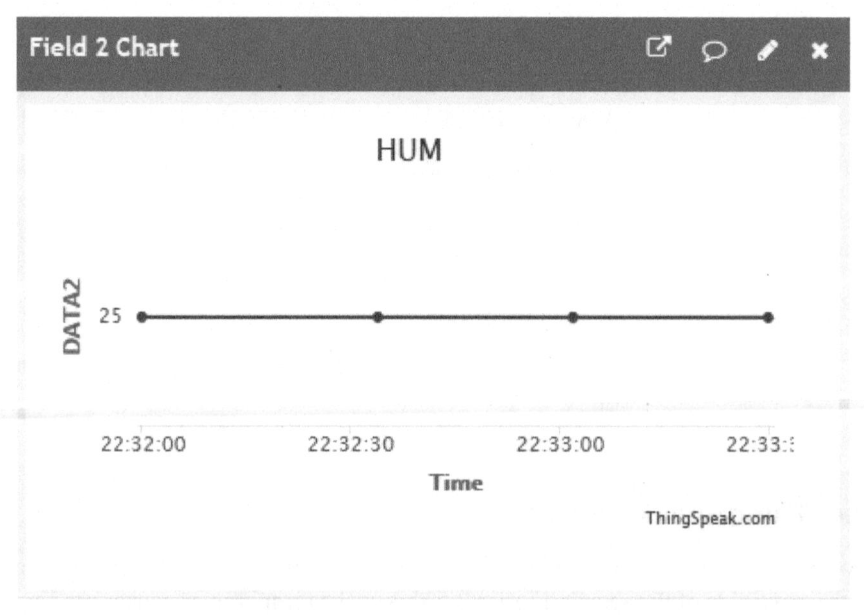

Figure 8.6 Field showing humidity sensor readings.

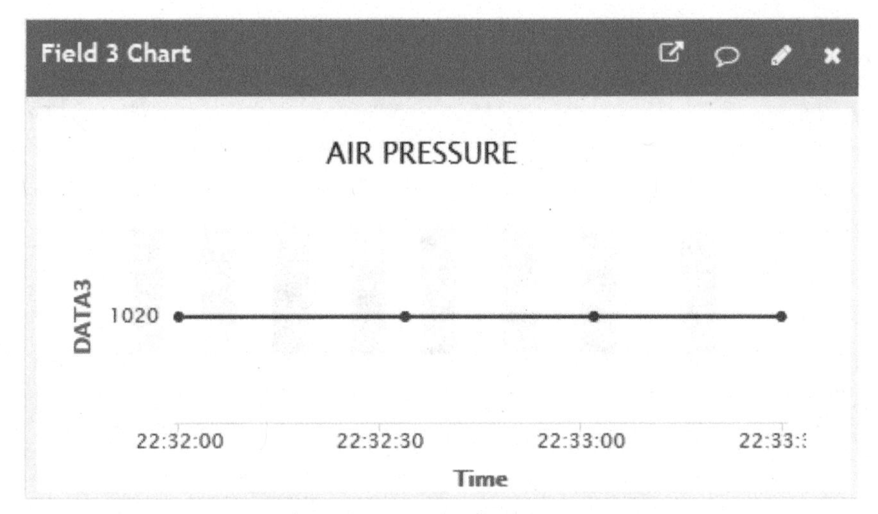

Figure 8.7 Field showing air pressure (mbar).

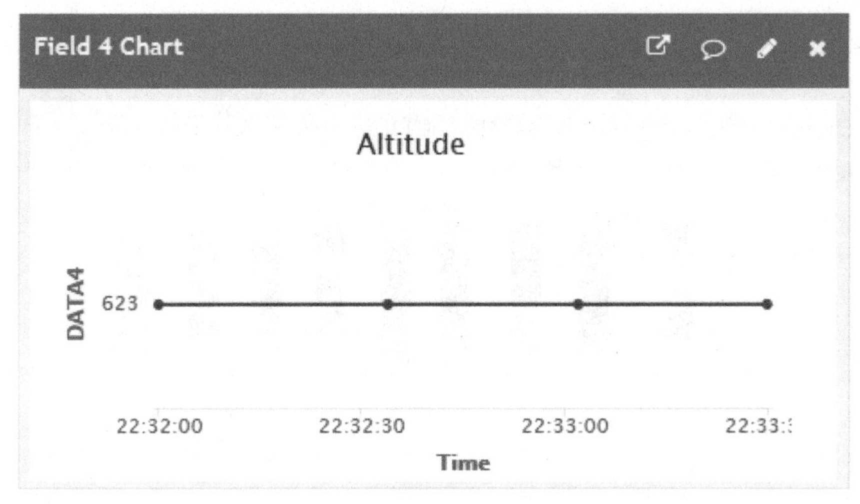

Figure 8.8 Field showing altitude.

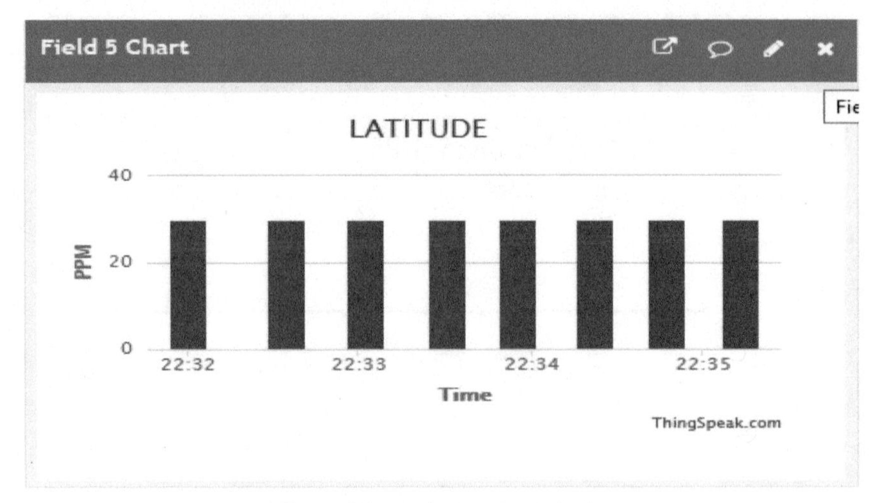

Figure 8.9 Field showing latitude.

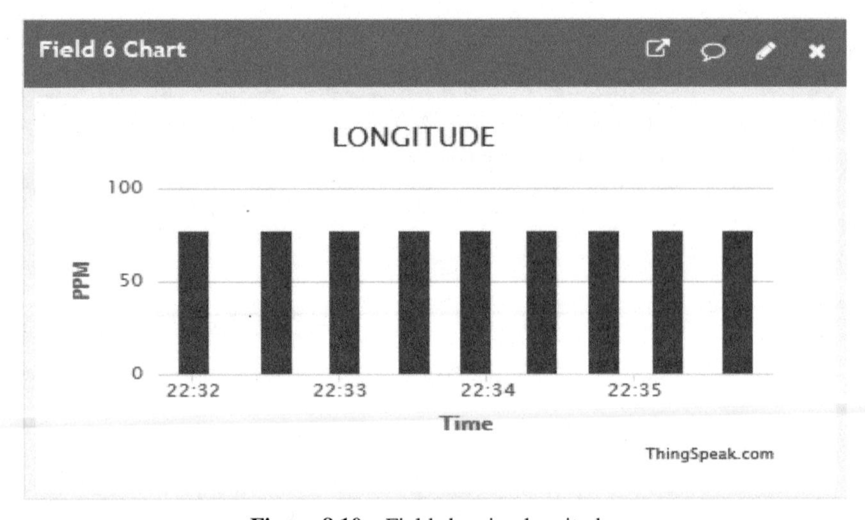

Figure 8.10 Field showing longitude.

9

Talking Road Unit at Pin Turn in Hilly Areas

Extra care is always required while driving the automotive vehicles in hilly regions and blind pin turns. Unless the driver not follows the driving etiquette on either side, occupants' safety is in danger. In this direction to address the identified issue, the system is proposed where sensors placed at both sides of the turn will talk to the vehicles passing through the area and alert the opposite vehicle driver during blind turning.

9.1 Introduction

Figure 9.1 shows the block diagram of the smart device1 with Ti Launch PAD, XBee, and other devices. It comprises a +12 V/500 mA power supply, a 12 to 5 V convertor, a Ti Launch PAD, an XBee, an LCD, a GPS, and IR/motion sensors 1 and 2. The main objective is to display the information regarding pin-turn road using IR/motion sensors 1 and 2 on the display unit by reading the sensors. The sensory information from the Ti Launch PAD is transferred to the XBee via a serial link to the other side of the pin-turn road.

Figure 9.2 shows the block diagram of smart device 2 with Ti Launch PAD, XBee, and other devices. It comprises a +12 V/500 mA power supply, a 12 to 5 V convertor, a Ti Launch PAD, an XBee, and an LCD. The main objective is to receive the information through the XBee modem and display the data of the pin-turn road in hilly areas using the display unit. Then the sensory information for the vehicle presence from the Ti Launch PAD is transferred to the NodeMCU modem via a serial link. The information is also uploaded to the cloud sever and mobile app (Blynk APP) using the NodeMCU.

Tables 9.1 and 9.2 show the list of the components to develop the system.

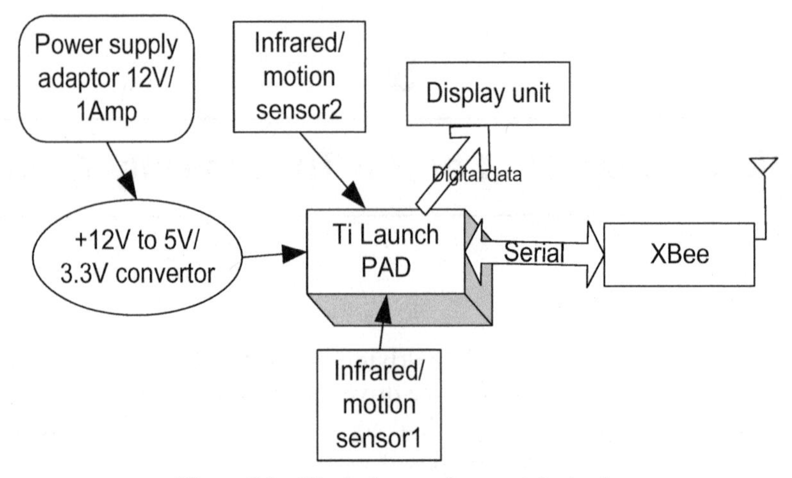

Figure 9.1 Block diagram for smart device 1.

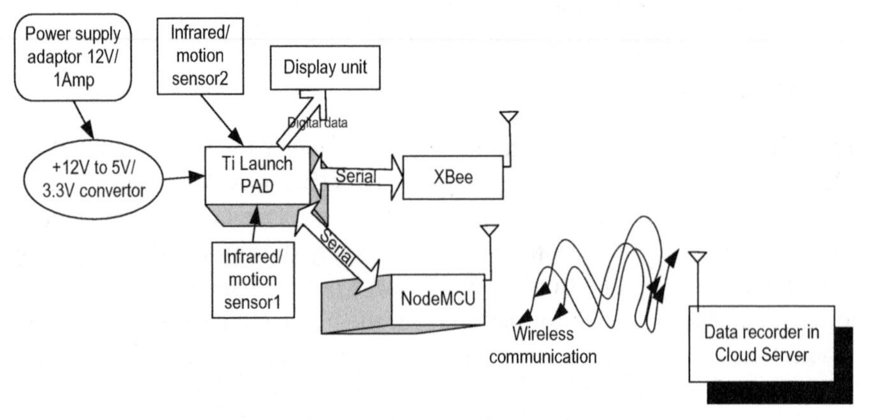

Figure 9.2 Block diagram for smart device 2.

9.2 Circuit Diagram

9.2.1 Circuit Diagram of Smart Device 1

1. GND pin of the power supply is connected to the GND pin of the Ti Launch PAD.
2. Pins 1 and 16 of the LCD are connected to GND of the power supply.
3. Pins 2 and 15 of the LCD are connected to +Vcc of the power supply.
4. Two fixed lags of the POT are connected to +5V and GND of the LCD and variable lag of the POT is connected to pin 3 of the LCD.

Table 9.1 Components' list for smart device 1

S.No	Components	Quantity
1	LCD20*4	1
2	LCD20*4 patch	1
3	DC 12 V/1 A adaptor	1
4	12 to 5 V, 3.3 V converter	1
5	LED with 330 Ω resistor	1
6	Jumper wire M to M	20
7	Jumper wire M to F	20
8	Jumper wire F to F	20
9	Motion sensor 1	1
10	Motion sensor 2	1
11	Ti Launch PAD	1
12	XBee	1
13	XBee breakout board	1

Table 9.2 Components' list of smart device 2

S.No	Components	Quantity
1	NodeMCU	1
2	LCD20*4	1
3	LCD20*4 patch	1
4	DC 12 V/1 A adaptor	1
5	12 to 5 V, 3.3 V converter	1
6	LED with 330 Ω resistor	1
7	Jumper wire M to M	20
8	Jumper wire M to F	20
9	Jumper wire F to F	20
10	Ti Launch PAD	1
11	XBee	1
12	XBee breakout board	1
13	Motion sensor 1	1
14	Motion sensor 2	1

5. RS, RW, and E pins of the LCD are connected to pins D1 = P2.0, GND, and D2 = P2.1 of the Ti Launch PAD.
6. D4, D5, D6, and D7 pins of the LCD are connected to pins D3 = P2.2, D4 = P2.3, D5 = P2.4, and D6 = P2.5 of the Ti Launch PAD.
7. +5V and GND pins of **motion sensor 1** are connected to +5V and GND pins of the power supply, respectively.
8. OUT1 of **motion sensor 1** is connected to P1_3 of the Ti Launch PAD.
9. +5V and GND pins of **motion sensor 2** are connected to +5V and GND pins of the power supply, respectively.
10. OUT1 of **motion sensor 2** is connected to P1_3 of the Ti Launch PAD.

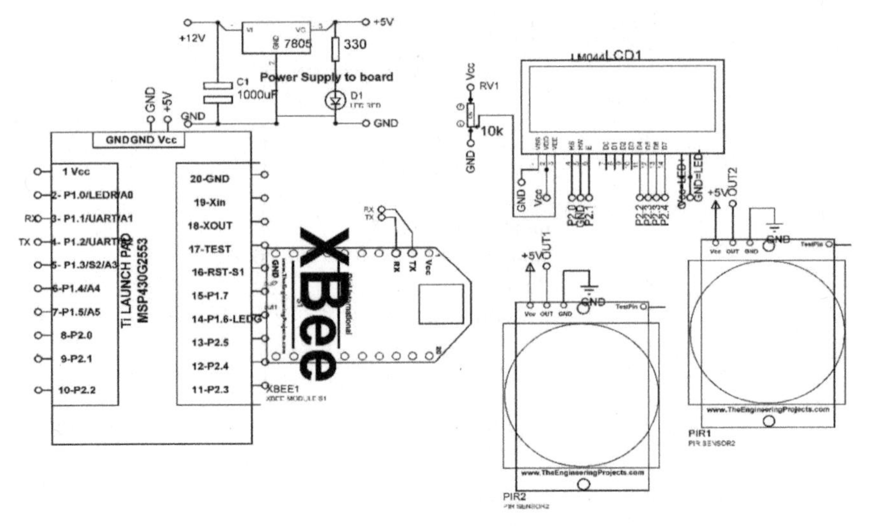

Figure 9.3 Circuit diagram of device 1.

11. Connect +Vcc, GND, and TX pins of the XBee to +5V, GND, and P1_1 pins of the Ti Launch PAD.

9.2.2 Circuit Diagram of Smart Device 2

1. +5V pin of the power supply is connected to the Vcc pin of the launch pad and WiFi modem/NodeMCU.
2. GND pin of the power supply is connected to the GND pin of the launch pad and the Ti Launch PAD.
3. Pins 1 and 16 of the LCD are connected to GND of the power supply.
4. Pins 2 and 15 of the LCD are connected to +Vcc of the power supply.
5. Two fixed lags of the POT are connected to +5V and GND of the LCD and variable lag of the POT is connected to pin 3 of the LCD.
6. RS, RW, and E pins of the LCD are connected to pins D1 = P2.0, GND, and D2 = P2.1 of the Ti Launch PAD.
7. D4, D5, D6, and D7 pins of the LCD are connected to pins D3 = P2.2, D4 = P2.3, D5 = P2.4, and D6 = P2.5 of the Ti Launch PAD.
8. +5V and GND pins of **motion sensor 1** are connected to +5V and GND pins of the power supply, respectively.
9. OUT1 of **motion sensor 1** is connected to P1_6 of the Ti Launch PAD.

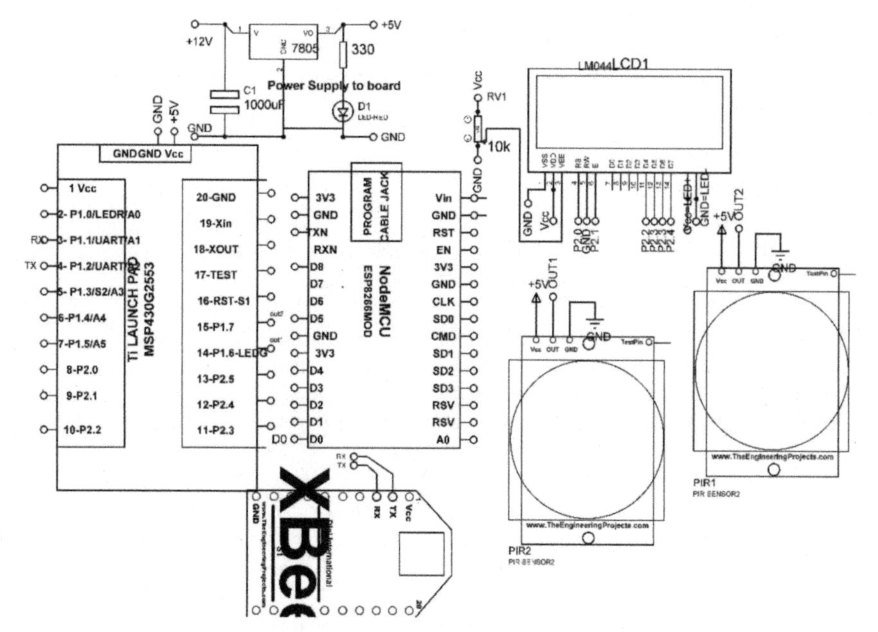

Figure 9.4 Circuit diagram of smart device 2.

10. +5V and GND pins of **motion sensor 2** are connected to +5V and GND pins of the power supply, respectively.
11. OUT1 of **motion sensor 2** is connected to P1_7 of the Ti Launch PAD.
12. Connect the P1.2 pin of the Ti Launch PAD to the RX pin of the NodeMCU.
13. Connect +Vcc, GND, and RX pins of the XBee to +5V, GND, and P1_2 pins of the Ti Launch PAD.

Figures 9.3 and 9.4 show the circuit diagram of smart devices 1 and 2, respectively.

9.3 Program Code

9.3.1 Code for Ti Launch Pad for Smart Device 1

```
#include <LiquidCrystal.h>
const int rs =P1_0, en = P1_1, d4 =P1_2, d5 = P1_3, d6 =P1_4,
     d7 = P1_5;
LiquidCrystal lcd(rs, en, d4, d5, d6, d7); // define pin of
    LCD for Ti launch pad
```

```
String inputStringXBEE_RX="";
int OTHER_SIDE_ACTUAL1,OTHER_SIDE_ACTUAL2; // define int
int pirPin1=P1_6;
int pirPin2=P1_7;
int HORN=P1_0;
void setup()
{
 Serial.begin(9600); // initialize serial communication
 lcd.begin(20, 4); // initialize LCD
 pinMode(pirPin1, INPUT_PULLUP); // set P1_6 as input
 pinMode(pirPin2, INPUT_PULLUP); // set P1_7 as input
 pinMode(HORN,OUTPUT); // set P1_0 as output
 lcd.setCursor(0,1); // set cursor of the LCD
 lcd.print("Pin TURN in HILLY area"); // print string on LCD
}
void loop()
{
 int  MOTION 1=digitalRead (pirPin1); // read PIR sensor1
 int  MOTION 2=digitalRead (pirPin2); // read PIR sensor2
 XBEE_RX (); // call function
 if((MOTION1==LOW)&&(MOTION2==LOW)) // check condition
 {
  int MOTION1=10; //assume int as 10
  int MOTION2=10; // assume int as 10
  Serial.print (MOTION1); // send serial value of motion
    sensor 1
  Serial.print (";"); // print string on serialport
  Serial.print (MOTION2); // send serial value of motion
    sensor 1
  Serial.print('\n'); // print new line character on serial
    port
  delay(10); // wait for 10 mSec
}
if((OTHER_SIDE_ACTUAL1==10)&&(OTHER_SIDE_ACTUAL1==10))
{
 lcd.setCursor (0,2); // set cursor on LCD
 lcd.print ("VECHILE PRESENT     "); // print string on LCD
 digitalWrite (HORN,HIGH); // make horn pin high
 }
  else
{
  int MOTION1=20; // assume int as 20
  int MOTION2=20; // assume int as 20
  lcd.setCursor(0,2); // set cursor on LCD
  lcd.print("VECHILE NOT PRESENT     "); // print string on
    LCD
```

```
  digitalWrite(HORN,LOW); // make horn pin low
  Serial.print(MOTION1); // print value on serial
  Serial.print(";");//  print string on serial
  Serial.print(MOTION2); // print value on serial
  Serial.print('\n'); // print new line character on serial
  delay(10); // wait for 10mSec
}

}


void XBEE_RX()
{
  while (Serial.available()>0)
  {
 inputStringXBEE_RX =Serial.readStringUntil('\n');// Get
    serial input
 OTHER_SIDE_ACTUAL1=(((inputStringXBEE_RX[0]-48)*10) + ((
    inputStringXBEE_RX[1]-48)*1));
 OTHER_SIDE_ACTUAL2=(((inputStringXBEE_RX[3]-48)*10) + ((
    inputStringXBEE_RX[4]-48)*1));;

   }
  inputStringXBEE_RX = ""; // make string empty
  delay(100); // wait for 100 mSec
  }
```

9.3.2 Program Code for Ti Launch Pad for Smart Device 2

```
#include <LiquidCrystal.h>
const int rs =P1_0, en = P1_1, d4 =P1_2, d5 = P1_3, d6 =P1_4,
     d7 = P1_5;
LiquidCrystal lcd(rs, en, d4, d5, d6, d7); // assign pins of
     LCD to Ti Launch pad


String inputStringXBEE_RX=""; // assign string
int OTHER_SIDE_ACTUAL1,OTHER_SIDE_ACTUAL2; // assume integer
int pirPin1=P1_6; // assign int to PIN1_6
int pirPin2=P1_7; // assign int to PIN1_7
int HORN=P1_0; // assign int to PIN1_0
void setup()
{
 Serial.begin (9600); // Initialize serial communication
 lcd.begin (20, 4); // Initialize LCD
 pinMode (pirPin1, INPUT_PULLUP); // assign pirPin1 as input
```

```
pinMode (pirPin2, INPUT_PULLUP); // assign pirPin1 as input
pinMode (HORN,OUTPUT); // assign HORN as output
lcd.setCursor(0,1); // set cursor of LCD
lcd.print("Pin TURN in HILLY area"); // print string on LCD
}
void loop()
{


 int  MOTION1=digitalRead(pirPin1); // read sensor 1
 int  MOTION2=digitalRead(pirPin2); // read sensor 2
 XBEE_RX();
 if((MOTION1==LOW)&&(MOTION2==LOW)) // Read condition
 {
  int MOTION1=10; // assign int value 10
  int MOTION2=10; // assign int value 10
  Serial.print(MOTION1); // print serial value
  Serial.print(";"); // print string on LCD
  Serial.print(MOTION2); // print serial value
  Serial.print('\n'); // print new line character on serial
  delay(10);
  }
  if((OTHER_SIDE_ACTUAL1==10)&&(OTHER_SIDE_ACTUAL1==10))
  {
   lcd.setCursor(0,2); // set cursor of LCD
   lcd.print("VECHILE PRESENT     "); // print string on LCD
   digitalWrite(HORN,HIGH); // make HORN pin HIGH
  }


  else
  {
   int MOTION1=20; // assign int as 20
   int MOTION2=20; // assign int as 20
   lcd.setCursor(0,2); // set cursor of LCD
   lcd.print("VECHILE NOT PRESENT     "); // print string on
    LCD
   digitalWrite(HORN,LOW); // make horn pin LOW
   Serial.print(MOTION1); // print serial value
   Serial.print(";"); // print string on LCD
   Serial.print(MOTION2); // print serial value
   Serial.print('\n'); // print new line char on serial
   delay(10); // wait for 10 mSec
  }
  }
```

```
void XBEE_RX() // function to read serial data
{
  while (Serial.available()>0) // check serial
  {
 inputStringXBEE_RX =Serial.readStringUntil('\n');// Get
    serial input
 OTHER_SIDE_ACTUAL1=(((inputStringXBEE_RX[0]-48)*10) + ((
    inputStringXBEE_RX[1]-48)*1));
 OTHER_SIDE_ACTUAL2=(((inputStringXBEE_RX[3]-48)*10) + ((
    inputStringXBEE_RX[4]-48)*1));;
  }
  inputStringXBEE_RX = ""; // make string empty
  delay(100); // wait for 100 mSec
  }
```

9.3.3 Program Code for Node MCU in Smart Device 2

```
#define BLYNK_PRINT Serial
#include <ESP8266WiFi.h>
#include <BlynkSimpleEsp8266.h>
///// add LCD library
#include <LiquidCrystal.h>
LiquidCrystal lcd(D0, D1, D2, D3, D4, D5);
///// add Wi-Fi credentials
char auth[] = "8507cac915f04a1bb4b00987e420afa0";
char ssid[] = "ESPServer_RAJ"; // hotspot ID
char pass[] = "RAJ@12345"; // hotspot password


BlynkTimer timer;
int HORN=D0; // assign HORN as pin D0
WidgetLCD blynkDISPLAY(V1); // assign LCD of blynk app as
    virtual pin V1
void READ_SENSOR() // function to read serial data from Ti
    Launch Pad
{
 XBEE_RX(); // call function
 if((OTHER_SIDE_ACTUAL1==10)&&(OTHER_SIDE_ACTUAL2==10)
 {
 blynkDISPLAY.print(0,1,"motion Detected     "); // print
    string on lCD
 digitalWrite(HORN,HIGH); // make HORN pin HIGH
 Blynk.virtualWrite(V2,10); // Write 10 on virtual pin V2
 delay(20); // wait for 20mSec
 }
```

```
 else

 {
 lcd.setCursor(0,1); // set cursor of LCD
 blynkDISPLAY.print(0,1,"motion NOT Detected"); // print
    string on LCD
 digitalWrite(HORN,LOW); // Make HORN pin LOW
 Blynk.virtualWrite(V2,20); // Write 20 on virtual pin V2
 delay(20); // wait for 20mSec
 }

}
void setup()
{
 Serial.begin(9600); // initialse serial
 lcd.begin(20, 4);  // initialize LCD
 pinMode(HORN,OUTPUT);/// assign HORN pin as output
 lcd.setCursor(0,0); // set cursor of LCD
 lcd.print("welcome"); // print string on LCD
 Blynk.begin(auth, ssid, pass); // initialize blynk APP
 timer.setInterval(10000L,READ_SENSOR);//// set sampling time
     to sample the READ_SENSOR ( ) function
 delay(2000); // wait for 2 Sec


}


void loop()
{
Blynk.run(); // initialse blynk APP
timer.run(); // Initiates BlynkTimer
}


void XBEE_RX()

{
  while (Serial.available()>0) // check serial
  {
    inputStringXBEE_RX =Serial.readStringUntil('\n');// Get
    serial input
    OTHER_SIDE_ACTUAL1=(((inputStringXBEE_RX[0]-48)*10) + ((
    inputStringXBEE_RX[1]-48)*1));
    OTHER_SIDE_ACTUAL2=(((inputStringXBEE_RX[3]-48)*10) + ((
    inputStringXBEE_RX[4]-48)*1));;
```

```
  }
  inputStringXBEE_RX = ""; // make string empty
  delay(100); // wait for 100 mSec
  }
}
```

9.4 BLYNK App

Follow the steps described in Section 4.4 to design the front end of the APP for the proposed system (Figure 9.5).

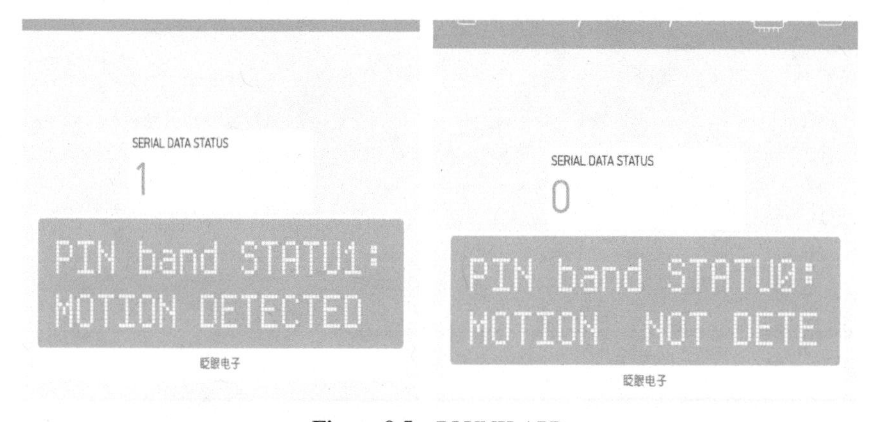

Figure 9.5 BLYNK APP.

10

Real-time Car Telematics Tracking System

A tracking device with own power supply that relies on sensors along with the GPS systems to gather data is the demand of the present scenario. A sensor-based device is proposed with a vehicle which can get the location as well as information on idle time, speed, and rash driving. The device is safe for use in mobility vehicles. The good thing about sensor-based devices is that one can easily get more information from the tracking system.

10.1 Introduction

The system can be designed in two different methods: one with ESP8266/NodeMCU and other with GPRS. Figure 10.1 shows the block diagram of the real-time car telematics tracking using NodeMCU and its associated external devices like LCD, keypads, and GPS. It comprises a +12 V/500 mA power supply, a 12 to 5 V convertor, a NodeMCU, an LCD, a keypad to enter the information, and a GPS. The main objective is to display the location information or coordinates on the LCD by reading the GPS and communicate it on the cloud.

Figure 10.2 shows the block diagram of the real-time car telematics tracking using GPRS modem and other devices. It comprises a +12 V/500 mA power supply, a 12 to 5 V convertor, GPRS as IoT modem, an LCD, a keypad to enter the data, and a GPS. The main objective is to display and update the location information or coordinates on the LCD by reading the GPS. The information is uploaded on the cloud server using the IoT modem.

Tables 10.1 and 10.2 show the components' list to develop the system.

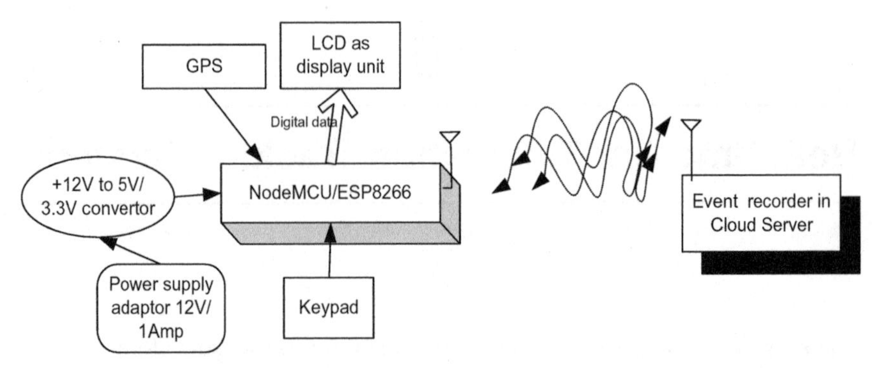

Figure 10.1 Block diagram of a smart device in cars using ESP8266.

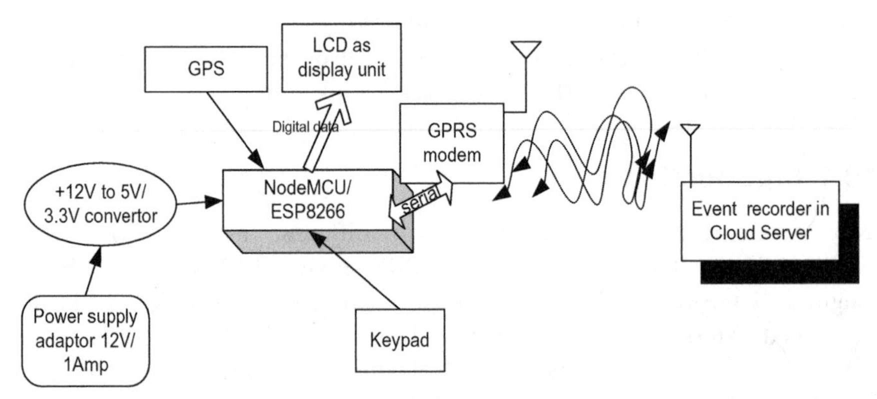

Figure 10.2 Block diagram of a smart device in cars using the GPRS modem.

Table 10.1 Components' list for the smart device using NodeMCU

S.No	Components	Quantity
1	GPS	1
2	LCD20*4	1
3	LCD20*4 patch	1
4	DC 12 V/1 A adaptor	1
5	12 to 5 V, 3.3 V converter	1
6	LED with 330 Ω resistor	1
7	Jumper wire M to M	20
8	Jumper wire M to F	20
9	Jumper wire F to F	20
10	NodeMCU	1
11	Four-switch keypad	1
12	Patch for NodeMCU	1

Table 10.2 Components' list for a smart device using GPRS

S.No	Components	Quantity
1	GPRS modem	1
2	LCD20*4	1
3	LCD20*4 patch	1
4	DC 12 V/1 A adaptor	1
5	12 to 5 V, 3.3 V converter	1
6	LED with 330 Ω resistor	1
7	Jumper wire M to M	20
8	Jumper wire M to F	20
9	Jumper wire F to F	20
10	NodeMCU	1
11	Four-key keypad	1
12	GPS	1
13	Patch for NodeMCU	1

10.2 Circuit Diagram

10.2.1 Connection of Smart Device Using NodeMCU/ESP8266

The following are the interfacing connections of the NodeMCU and other devices.

1. +5V and GND pins of the NodeMCU are connected to +5V and GND pins of the power supply.
2. Pins 1 and 16 of the LCD are connected to GND of the power supply.
3. Pins 2 and 15 of the LCD are connected to +5V of the power supply.
4. Fixed legs of the 10K POT are connected to +5V and GND of the power supply and variable leg to pin 3 of the LCD.
5. Pin D1, GND, and pin D2 of the NodeMCU are connected to pin 4(RS), pin 5(RW), and pin 6(E) of the LCD.
6. Pin D3, pin D4, pin D5, and pin D6 of the NodeMCU are connected to pin 11(D4), pin 12(D5), pin 13(D6), and pin 14(D7) of the LCD, repectively.
7. +Vcc, GND, and serial-out pins of the GPS are connected to +5V, GND, and RX pins of the NodeMCU, repectively.
8. Keypad switches are connected to D0, D7, D8, and D9 pins of the NodeMCU. Figure 10.3 shows the circuit diagram of a smart device using NodeMCU.

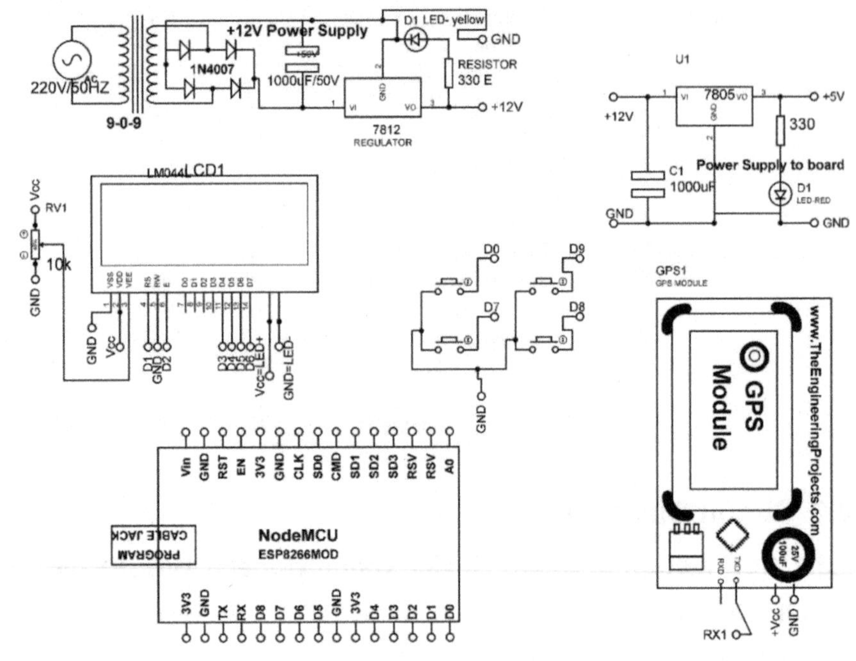

Figure 10.3 Circuit diagram of a smart device using NodeMCU/ESP8266.

10.2.2 Connection of Smart Device Using GPRS Modem

The following are the interfacing connections of the NodeMCU and the external devices.

1. +5V and GND pins of the NodeMCU are connected to +5V and GND pins of the power supply.
2. Pins 1 and 16 of the LCD are connected to GND of the power supply.
3. Pins 2 and 15 of the LCD are connected to +5V of the power supply.
4. Fixed legs of the 10K POT are connected to +5V and GND of the power supply and variable leg to pin 3 of the LCD.
5. Pin D1, GND, and pin D2 of the NodeMCU are connected to pin 4(RS), pin 5(RW), and pin 6(E) of the LCD.
6. Pin D3, pin D4, pin D5, and pin D6 of the NodeMCU are connected to pin 11(D4), pin 12(D5), pin 13(D6), and pin 14(D7) of the LCD.
7. +Vcc, GND, and serial-out pins of the GPS are connected to +5V, GND, and RX pins of the NodeMCU.
8. Keypad switches are connected to D0, D7, D8, and D9 pin of the NodeMCU.

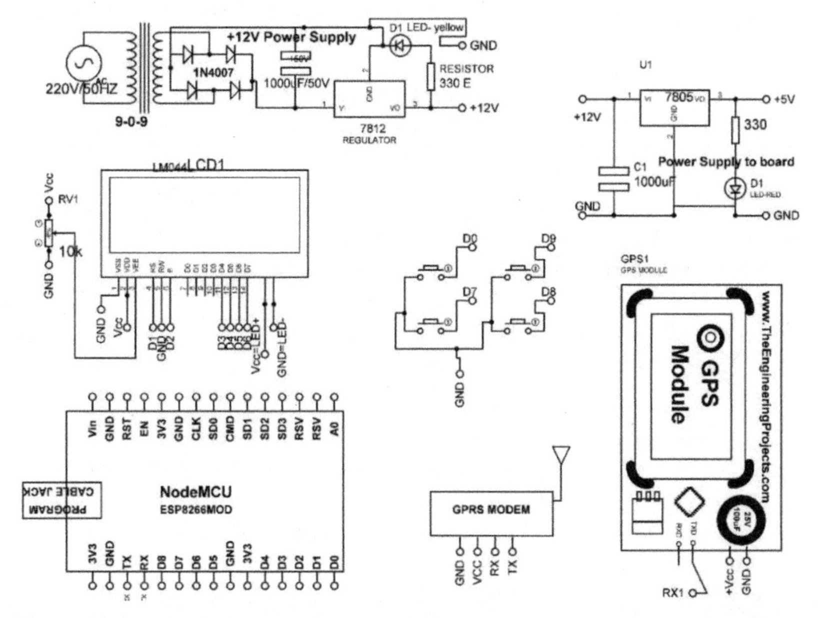

Figure 10.4 Circuit diagram of a smart device using NodeMCU and GPRS modem.

9. +Vcc, GND, and RX pins of the GPRS modem are connected to +5V, GND, and TX pins of the NodeMCU. Figure 10.4 shows the circuit diagram of a smart device using NodeMCU and GPRS modem.

10.3 Program Code

10.3.1 Program Code for Smart Device Using NodeMCU

```
#define BLYNK_PRINT Serial
#include <ESP8266WiFi.h> // attach ESP8266 library
#include <BlynkSimpleEsp8266.h> // attach blynk library
#include <LiquidCrystal.h> //  add LCD library
LiquidCrystal lcd(D0, D1, D2, D3, D4, D5);
#include <TinyGPS.h> // attach GPS library
TinyGPS gps;
void getgps(TinyGPS &gps);
float latitude, longitude; // assign latitude and longitude
   as float
byte a;
WidgetLCD LCD_BLYNK(V0); // assign virtual pin 0 to Blynk LCD
```

```
///// add credentials
char auth[] = "8507cac915f04a1bb4b00987e420afa0"; // token
char ssid[] = "ESPServer_RAJ"; // Hot spot ID
char pass[] = "RAJ@12345"; // hot spot password


BlynkTimer timer;
int BUTTON_ONE=D0; // D0 as int
int BUTTON_TWO=D7; // D7 as int
int BUTTON_THREE=D8; // D8 as int
int BUTTON_FOUR=D9; // D9 as int


void READ_SENSOR()
{
 CALL_GPS(); // call GPS
 Blynk.virtualWrite(V3,latitude); // write latitude on
    virtual pin V3 on blynk LCD
 Blynk.virtualWrite(V4,longitude); // write latitude on
    virtual pin V3 on blynk LCD
 int ONE=digitalRead(BUTTON_ONE); // read Button
 int TWO=digitalRead(BUTTON_TWO); // read Button
 int THREE=digitalRead(BUTTON_THREE); // read Button
 int FOUR=digitalRead(BUTTON_FOUR); // read Button
 if(ONE==LOW) // check condition
 {
 lcd.setCursor(0,1);  // set cursor on LCD
 lcd.print("EMERGENCY L1"); // print string on LCD
 LCD_BLYNK.print(0,1"EMERGENCY L1"); // print string on blynk
     LCD
 Blynk.virtualWrite(V5,ONE); // write value on virtual pin5
 delay(20); // wait for 10 mSec
 }
 if(TWO==LOW) // // check condition
 {
 lcd.setCursor(0,1); // set cursor on LCD
 lcd.print("EMERGENCY L2"); // print string on LCD
 LCD_BLYNK.print(0,1"EMERGENCY L2"); // print string on blynk
     LCD
 Blynk.virtualWrite(V5,TWO); // write on V5
 delay(20);// wait for 20 mSec
 }
  if(THREE==LOW) // check condition
  {
  lcd.setCursor(0,1); // set cursor on LCD
```

```
  lcd.print("EMERGENCY L3"); // print string on LCD
 LCD_BLYNK.print(0,1"EMERGENCY L3"); // print string on blynk
     LCD
  Blynk.virtualWrite(V5,THREE); // write on V5
  delay(20); // wait for 20 mSec
  }
   if(FOUR==3) // check condition
  {
  lcd.setCursor(0,1); // set cursor on LCD
  lcd.print("EMERGENCY L4"); // print string on LCD
 LCD_BLYNK.print(0,1"EMERGENCY L4"); // print string on blynk
     LCD
  Blynk.virtualWrite(V5,FOUR); // write on V5
  delay(20); // wait for 20 mSec
  }
  }


void setup()
{
 Serial.begin(9600); // for GPS
 lcd.begin(20, 4);
 pinMode(BUTTON_ONE,INPUT); // set pin as input
 pinMode(BUTTON_TWO,INPUT); // set pin as input
 pinMode(BUTTON_THREE,INPUT); // set pin as input
 pinMode(BUTTON_FOUR,INPUT); // set pin as input
 lcd.setCursor(0,0); // set cursor on LCD
 lcd.print("Car Telematics Sys."); // print string on LCd
 Blynk.begin(auth, ssid, pass); // initialse blynk
 timer.setInterval(10000L,READ_SENSOR);//// change
 delay(3000); // wait for 3000 mSec
 lcd.clear(); // clear the contents of LCD
}


void loop()
{
Blynk.run(); // run BLYNK
timer.run(); // Initiates BlynkTimer
}
 void getgps(TinyGPS &gps)
{
 float latitude, longitude;
 decode and display position data
 gps.f_get_position(&latitude, &longitude);
 lcd.setCursor(0,3); // set cursor on LCD
```

```
 lcd.print("Lat:"); // print string on LCD
 lcd.print(latitude,5); // print latitude value
 lcd.print(" "); // print string on LCD
 lcd.setCursor(10,3); // set cursor on LCD
 lcd.print("Long:"); //  print string on LCD
 lcd.print(longitude,5); // print longitude
 lcd.print(" "); // print string on LCD
 delay(3000); // wait for 3000 seconds
}


void  CALL_GPS()
{
 byte a;
 if (Serial.available() > 0) // if there is data coming into
    the serial line
 {
  a = Serial.read(); // get the byte of data
   if(gps.encode(a)) // if there is valid GPS data...
     {
       getgps(gps); // get the data and display it on the LCD
     }
   }

   }
```

10.3.2 Program Code for GPRS

```
#include <SoftwareSerial.h>
#include <String.h>
SoftwareSerial MyGPRS(D7,D8);
char thingSpeakAddress[] = "api.thingspeak.com";
//////////////// library for LCD
#include <LiquidCrystal.h>
LiquidCrystal lcd(P2_0, P2_1,P2_2, P2_3, P2_4, P2_5); //
    assign pin of LCD to Ti launch pad
//int8_t answer;
float answer;
#include <TinyGPS.h>
TinyGPS gps;
void getgps(TinyGPS &gps);
float latitude, longitude; // assume latitude and longitude
    as float
byte a; // assign a as byte
```

```
///// for keypad
int BUTTON_ONE=D0; // assign D0 as int
int BUTTON_TWO=D7; // assign D7 as int
int BUTTON_THREE=D8; // assign D8 as int
int BUTTON_FOUR=D9; // assign D9 as int
void CallGPRS() // function for GPS
{
 gprspwr_on();
 CALL_GPS(); // call function
 //connect gprs to internet
 answer = sendATcommand("AT+CGATT?","OK",5,2000); // send AT
    command
 answer = sendATcommand("AT+CSTT=\"CMNET\"","OK",3,2000); //
    send AT command
 answer = sendATcommand("AT+CIICR","OK",3,2000);// send AT
    command
 answer = sendATcommand("AT+CIFSR","OK",3,2000); // send AT
    command
 answer = sendATcommand("AT+CIPSPRT=0","OK",3,2000); // send
    AT command
 answer = sendATcommand("AT+CIPSTART=\"tcp\",\"api.
          thingspeak.com\",\"80\"","CONNECT OK",5,2000);
 int ONE=digitalRead(BUTTON_ONE); // read input
 int TWO=digitalRead(BUTTON_TWO); //read input
 int THREE=digitalRead(BUTTON_THREE); // read input
 int FOUR=digitalRead(BUTTON_FOUR); // read input
int param1=ONE;
int param2=TWO;
int param3=THREE;
int param4=FOUR;
int param5=latitude;
int param6=longitude;

  answer = senddata1(param1,param2,param3,param4,param5,
    param6);
  delay(3000);
  gprspwr_off();

  //put arduino to sleep?
  for (int i=0; i<60; i++)
  {
   delay(150);
  }
  }
  /**************************************gprs 2nd function
```

```
  end************************************/
void setup()
{
 // put your setup code here, to run once:
 MyGPRS.begin(9600);                    // the GPRS baud rate
 Serial.begin(9600);                      // the computer serial
    interface baud rate
 lcd.begin(20, 4);
 delay(1000);
 lcd.print("GPRS BASED IoT");
 delay(1000);
 pinMode(BUTTON_ONE,INPUT); // assign pin as input
 pinMode(BUTTON_TWO,INPUT); // assign pin as input
 pinMode(BUTTON_THREE,INPUT); //assign pin as input
 pinMode(BUTTON_FOUR,INPUT); // assign pin as input
}


void loop()
{
 byte l;
 serialEvent_NODEMCU(); // call function
 CallGPRS();    // call GPRS function
 delay(500);

}


/*******************************************/

int8_t senddata1(int data,int data1,int data2,int data3,
    int data4,int data5,int data6,int data7)

{


 MyGPRS.println("AT+CIPSEND");
 while(MyGPRS.available() > 0) MyGPRS.read();     // Clean the
    input buffer
 delay(500);
 MyGPRS.println("POST /update HTTP/1.1");     // Send the AT
   command
 while(MyGPRS.available() > 0) MyGPRS.read();     // Clean the
    input buffer
 delay(500);
```

```
MyGPRS.println("Host: api.thingspeak.com");    // Send the
   AT command
while(MyGPRS.available() > 0) MyGPRS.read();    // Clean the
    input buffer
delay(500);
MyGPRS.println("Connection: close");    // Send the AT
   command

while(MyGPRS.available() > 0)MyGPRS.read();    // Clean the
   input buffer
delay(500);
MyGPRS.println("X-THINGSPEAKAPIKEY:L5I8F6JM3NKUQNTU");//
   T1GIUPBKKRDPMWRX");
while(Serial2.available() > 0) Serial2.read();    // Clean
   the  input buffer
delay(500);

MyGPRS.println("Content-Type: application/x-www-form-
   urlencoded");    // Send the AT command


while(MyGPRS.available() > 0) MyGPRS.read();    // Clean the
    input buffer
delay(500);

MyGPRS.println("Content-Length:92");    // Send the AT
   command

while(MyGPRS.available() > 0) MyGPRS.read();    // Clean the
    input buffer
delay(500);

MyGPRS.println("");    // Send the AT command

while( MyGPRS.available() > 0) MyGPRS.read();    // Clean
   the input buffer
delay(500);
MyGPRS.print("&field1=");    // Send the AT command
MyGPRS.print(data);
MyGPRS.print("&field2=");    // Send the AT command
MyGPRS.print(data1);
MyGPRS.print("&field3=");    // Send the AT command
MyGPRS.print(data2);
MyGPRS.print("&field4=");    // Send the AT command
MyGPRS.print(data3);
MyGPRS.print("&field5=");    // Send the AT command
```

```
MyGPRS.print(data4);
MyGPRS.print("&field6=");     // Send the AT command
MyGPRS.print(data5);
MyGPRS.print("&field7=");     // Send the AT command
MyGPRS.print(data6);
MyGPRS.print("&field8=");     // Send the AT command
MyGPRS.print(data7);


while(MyGPRS.available() > 0) MyGPRS.read();     // Clean the
    input buffer
delay(500);
MyGPRS.println((char)26);
delay(500);
while(MyGPRS.available() > 0) MyGPRS.read();     // Clean the
    input buffer
delay(500);



  answer = 0;
  return answer;


}


void gprspwr_on()
{
 pinMode(5, OUTPUT);  // set pin as output
 digitalWrite(5,LOW); // make pin5 to LOW
 delay(1000); // wait for 1000 mSec
 digitalWrite(5,HIGH); // make pin5 to LOW
 delay(2000); // wait for 1000 mSec
 digitalWrite(5,LOW); // make pin5 to LOW
 readATcommand("Call Ready",6,10000);
 if (answer == 1)
  {


  }

}


void gprspwr_off()
```

```
{
 pinMode(5, OUTPUT);  // set pin 5 as output
 digitalWrite(5,LOW); // make pin 5 to LOW
 delay(1000); // wait for 1000 mSec
 digitalWrite(5,HIGH); // make pin5 to HIGH
 delay(2000);·// wait for 1000 mSec
 digitalWrite(5,LOW); // Make pin 5 to LOW
 answer = readATcommand("NORMAL POWER DOWN",2,2000);
 if (answer == 1)
 {


 }

}


boolean gprspwr_status()
{
 answer = sendATcommand("AT", "OK", 2, 2000);
 if (answer == 0)
  {


  }
  else if (answer == 1)
  {

  }
  return answer;
}


int8_t readATcommand(char* expected_answer1, unsigned int
    expected_answers, unsigned int timeout)

{
  uint8_t x = 0,  answer = 0;
  boolean complete = 0;
  char a;
  char response[100];
  unsigned long previous;
  String  incomingdata;
  boolean first;
```

```
previous = millis();
for(int i = 0; i < expected_answers; i++)
{
 x = 0;
 complete = 0;
 a = 0;
 first = 0;
 memset(response, '\0', 100);     // Initialize the string
 do
 {
  if(MyGPRS.available() != 0)
  {
   a = MyGPRS.read();
   //Serial.println(a,DEC);
   if (a == 13)
   {
    a = MyGPRS.read();
    //Serial.println(a,DEC);
    if (a == 10)
    {
     if (first == 0)
     {
      //keep going, just ignore it
     }
     else
     {
      complete = 1;
     }
    }
   }
   else if(a == 0)
   {
   }
   else
   {
    response[x] = a;
    x++;
    first = 1;
   }
   if(strstr(response, expected_answer1) != NULL)
   {
    answer = 1;
    complete = 1;
    return answer;
   }
```

```
    else if(strstr(response, "ERROR") != NULL)
    {
     answer = 2;
    }
   }
 }
 while((complete == 0) && ((millis() - previous) < timeout));


  }

  return answer;
}


int8_t sendATcommand(char* ATcommand, char* expected_answer1,
    unsigned int expected_answers, unsigned int timeout)
{

 uint8_t x=0,  answer=0;
 boolean complete = 0, first = 0;
 char a;
 char response[100];
 unsigned long previous;
 String  incomingdata;
 delay(100);  // wait for 100 mSec
 //Serial.println("Send AT Command");
 while(MyGPRS.available() > 0) MyGPRS.read();    // Clean the
    input buffer
 MyGPRS.println(ATcommand);    // Send the AT command
 previous = millis();
 for(int i = 0; i < expected_answers; i++)
{
  x = 0;
  complete = 0;
  a = 0;
  first = 0;
  memset(response, '\0', 100);    // Initialize the string
  do
{
  if (MyGPRS.available() != 0)
   {
     a = MyGPRS.read(); // read GPRS
     if (a == 13)
     {
     a = MyGPRS.read();
```

```
        if (a == 10){
          if (first == 0)
          {
           //keep going, just ignore it
          }
          else
          {
           complete = 1;
          }
         }
        }
        else if(a == 0)
        {

         }
        else
        {
         response[x] = a;
         x++;
         first = 1;

       }
       if (strstr(response, expected_answer1)!=NULL)

       {
        answer = 1;
        complete = 1;
       }
       else if(strstr(response, "ERROR")!=NULL)
       {
        answer = 2;
        complete = 1;
       }
      }
    }
    while((complete == 0) && ((millis() - previous) < timeout));


    }
    return answer;
  }
   void getgps(TinyGPS&gps)
   {
    float latitude, longitude;
    decode and display position data
    gps.f_get_position(&latitude, &longitude);
```

```
    lcd.setCursor(0,3); // set cursor on LCD
   lcd.print("Lat:"); // print string on LCD
   lcd.print(latitude,5); // print value of latitude value
   lcd.print(" "); // print string on LCD
    lcd.setCursor(10,3);// set cursor on LCD
   lcd.print("Long:"); // print string on LCD
   lcd.print(longitude,5); // // print value of longitude
     value
   lcd.print(" "); // print string on LCD
   delay(3000); // wait for 3 seconds
}

void  CALL_GPS()
{
 byte a;
 if (Serial.available() > 0) // if there is data coming into
    the serial line
 {
  a = Serial.read(); // get the byte of data
  if(gps.encode(a)) // if there is valid GPS data...
      {
         getgps(gps); // get the data and display it on the
    LCD
      }
 }


 }
```

10.4 BLYNK App

Follow the steps described in Section 4.4 for BLYNK app development and design front end of APP for the proposed system (Figure 10.5).

10.5 ThingSpeak Server

Follow the steps described in Section 5.4 for ThingSpeak server development and design front end of the APP for the proposed system, and check the data on different fields (Figure 10.5). Figure 10.6 shows the longitude, Figure 10.6 shows latitude.

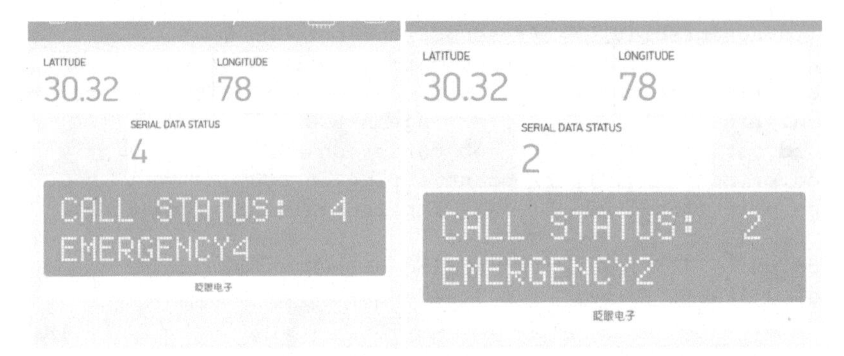

Figure 10.5 BLYNK APP.

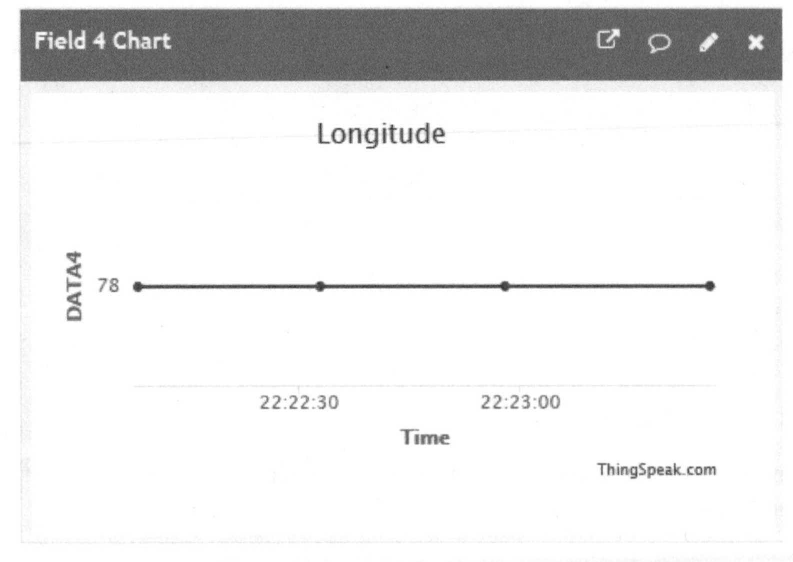

Figure 10.6 Field showing longitude.

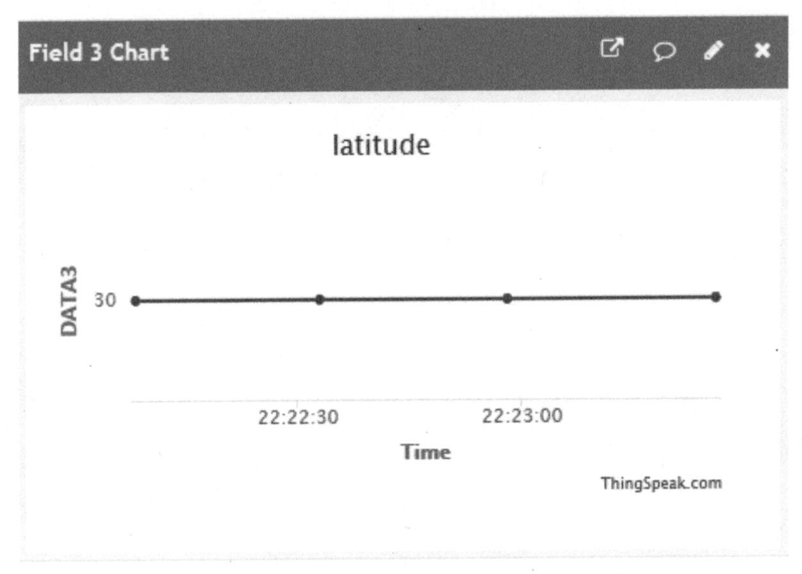

Figure 10.7 Field showing latitude.

References

[1] http://saphanatutorial.com/introduction-to-internet-of-things-part-1/
[2] https://www.postscapes.com/internet-of-things-protocols/
[3] http://www.c-sharpcorner.com/UploadFile/f88748/internet-of-things-part-2/
[4] Sethi, P., and Sarangi, S. R. (2017). Internet of things: architectures, protocols, and applications. *J. Elec. Comput. Eng.* 2017.
[5] http://ieeexplore.ieee.org/stamp/stamp.jsp?tp=&arnumber=7805273
[6] https://internetofthingswiki.com/iot-applications-examples/541/#Agriculture
[7] http://www.libelium.com/resources/top_50_iot_sensor_applications_ranking/
[8] https://ac.els-cdn.com/S2314728816300149/1-s2.0-S2314728816300149-main.pdf?_tid=b8305708-1482-11e8-97ae-00000aacb35e&acdnat=1518941418_2cb9b74701d8abb1b4a8a9ed8c643a7c- IoT PLATFORM (upto 27)
[9] https://internetofthingswiki.com/top-20-iot-platforms/ IoT platform
[10] Evens, D. (2011). The Internet of Things How the Next Evolution of the Internet is Changing Everything. Cisco Internet Business Solutions Group (IBSG). CISCO White Paper, 1(2011), 1–11.
[11] CII (2017). Technological disruption and the automotive industry – Automotive Industry 4.0 Summit “Connected & Intelligent” Roland Berger, 32.
[12] https://www.amazon.in/internet-Things-Hands-Arsheep-Bahga/dp/8173719543

Index

About the Authors

Dr. Rajesh Singh is currently associated with Lovely Professional University as Professor with more than fifteen years of experience in academics. He has been awarded as gold medalist in M.Tech and honors in his B.E. His area of expertise includes embedded systems, robotics, wireless sensor networks and Internet of Things. He has organized and conducted a number of workshops, summer internships and expert lectures for students as well as faculty. He has twenty three **patents** in his account. He has published around hundred **research papers** in referred journals/conferences.

Under his mentorship students have participated in national/international competitions including Texas competition in Delhi and Laureate award of excellence in robotics engineering in Spain. Twice in last four years he has been awarded with **"certificate of appreciation"** and **"Best Researcher award-2017"** from University of Petroleum and Energy Studies for exemplary work. He got **"certificate of appreciation"** for mentoring the projects submitted to Texas Instruments Innovation challenge India design contest, from Texas Instruments, in 2015. He has been honored with young investigator award at the International Conference on Science and Information in 2012. He has published ten books in the area of Embedded Systems and Internet of Things with reputed publishers like CRC/Taylor & Francis, Narosa, GBS, IRP, NIPA and RI publication. He is editor to a special issue published by AISC book series, Springer with title "Intelligent Communication, Control and Devices"-2017 & 2018.

Dr. Anita Gehlot is associated with Lovely Professional University as Associate Professor with more than ten years of experience in academics. She has **twenty patents** in her account. She has published more than **fifty research papers** in referred journals and conference. She has organized a number of workshops, summer internships and expert lectures for students. She has been awarded with **"certificate of appreciation"** from University of Petroleum and Energy Studies for exemplary work. She has published ten books in the area of Embedded Systems and Internet of Things with reputed publishers like CRC/Taylor & Francis, Narosa, GBS, IRP, NIPA and RI publication. She is editor to a special issue published by AISC book series, Springer with title "Intelligent Communication, Control and Devices-2018".

Dr. Raghuveer Chimata is currently working as a Postdoctoral Appointee at Argonne National Laboratory (ANL) working on Quantum Materials Simulations. He completed his Ph.D in Theoretical Material Science mainly into Spin dynamics, Ultrafast remagnetization dynamics and studying magnetic properties of Spin Chains and spinels from Uppsala University, Swedan. And his PG in Master of Computational Sciences in the area of Scientific Computing and Solid State Theory, Uppsala University, Swedan. His research contributions are published in very reputed International Journals like Physical Review Letters, Computational Material Sciences, Physical Reviews B 92 etc...and many more. His research interests are: Density Functional Theory (DFT), Hybrid Functionals, Electronic structure calculations, Magnetic and optical properties of d and f electron systems using DFT based tools. Spin Dynamics, Magnetic properties of spinels and spin chain compounds, Ultrafast remagnetization, magnonics, complex magnetic oxides, magnetic amorphous RE-TE materials and Ultrafast switching.

Bhupendra Singh is Managing director of Schematics Microelectronics and provides Product design and R&D support to industries and Universities. He has completed BCA, PGDCA, M.Sc. (CS), M.Tech and has more than eleven years of experience in the field of Computer Networking and Embedded systems. He has published ten books in the area of Embedded Systems and Internet of Things with reputed publishers like CRC/Taylor & Francis, Narosa, GBS, IRP, NIPA and RI publication.

Dr. P. S. Ranjit is currently associated with Aditya Engineering College, India. He has so far published more than 34 Research Papers in National and International Journals and Conferences and published 10 patents in Automotive Technology. He completed his Ph.D in Internal Combustion engines in the area of Vegetable oils and Hydrogen supplementation from UPES, Dehradun, M.E. in Mechanical Engineering from University of Madras

B.E. in Automobile Engineering from Shivaji University, Kolhapur with University Second rank and 3 years Diploma in Automobile Engineering form State Board of Technical Education, Hyderabad.

Dr. P. S. Ranjit actively involved in sponsored research projects. Very recently 8.1 Million Indian Rupees financially sponsored from Ministry of New and Renewable Energy (MNRE), Government of India completed in the area of Hydrogen utilization in SVO based IC Engines. He developed the Engine Research Laboratory at UPES with NFPA Class I, Division 2, Group B standards in order to store and handling the ultra-high purity gaseous Hydrogen.